PRAISE FOR CHURC

"Leaming has produced a splendid book about Britain's last lion. Even in winter, he never lost his roar." —*Washington Times*

"Insightful and highly entertaining. . . . Leaming's masterfully written and intimate book is a rich insider's tale of backstairs palace intrigue. Her superb sources provide wonderful insight into the personalities of the players and Leaming brings them alive in this period as few other authors have done."

—*Sarasota Herald-Tribune*

"Startling. . . . An absorbing and detailed account."

—*New Statesman* magazine (UK)

"Fascinating. . . . A memorable and very readable portrait of one of the twentieth century's towering figures."

—*Roanoke Times* (Virginia)

"The plot, which covers the last ten years of [Churchill's] political life, could hardly be more gripping. . . . Barbara Leaming has fashioned a poignant narrative. Like any good storyteller, she focuses sharply on the protagonist and his aspirations. . . . Compelling." —*Sunday Times* (London)

"Leaming spins a gripping yarn of epic dimensions. . . . A moving chronicle of one extraordinary man's determination to overcome seemingly crushing adversity. . . . In the true sense of the word, it is a great story. . . . A fascinating take on momentous events, with a truly human hero at its heart." —*Eastern Daily Press* (UK)

"Sympathetic and highly readable. . . . Leaming takes the reader on an intimate cruise of British politics." —*San Francisco Chronicle*

"Colorful details abound. . . . A beguiling portrait of a leading politician who refuses to acknowledge his waning powers."

—*Daily Express* (London)

"Leaming tells a rousing story. . . . This is exciting historical writing. . . . Churchill was the greatest figure of an epic century. You need to know about the last ten years of his public life."

—Liz Smith

"Accomplished biographer Leaming tracks Winston Churchill's masterful postwar comeback. . . . A lively chronicle. . . . Tight, polished, and effectively focused on the lesser-known end of Churchill's career." —*Kirkus Reviews*

"Absorbing. . . . Illuminating. . . . Eminently readable political history." —*Booklist*

"Magnificent. . . . A compelling, vivid, and deeply poignant portrait. . . . Highly recommended." —*Catholic Herald* (London)

"Fascinating insight into the workings of a master politician and statesman." —*Daily News Online* (Washington)

"A well-written book. . . . Full of interesting details. . . . The manner in which Leaming has captured the essential Winston commands our admiration." —*Church Times* (London)

ABOUT THE AUTHOR

Barbara Leaming is the author of two *New York Times* bestselling biographies and three *New York Times* Notable Books of the Year. Her most recent book, a biography of John F. Kennedy, focused on the influence of British history and culture on the thirty-fifth president. She was the first to write extensively about the extraordinary influence of Winston Churchill on Kennedy's intellectual formation and political strategies. Her articles have appeared in many publications in the US and Europe, including the *New York Times Magazine*, *Vanity Fair*, and *The Times* (London). She is married and lives in Connecticut.

CHURCHILL DEFIANT

CHURCHILL DEFIANT

FIGHTING ON

1945–1955

BARBARA LEAMING

HARPER PERENNIAL

NEW YORK • LONDON • TORONTO • SYDNEY • NEW DELHI • AUCKLAND

HARPER PERENNIAL

A hardcover edition of this book was published in 2010 by HarperCollins Publishers.

Frontispiece courtesy of Toni Frisell/Getty Images.

FIRST HARPER PERENNIAL EDITION PUBLISHED 2011.

Designed by William Ruoto

Library of Congress Cataloging-in-Publication Data has been applied for.

ISBN 978-0-06-133760-4 (pbk.)

11 12 13 14 15 ID/RRD 10 9 8 7 6 5 4 3 2 1

CONTENTS

CHURCHILL DEFIANT

I

YOU WILL, BUT I SHALL NOT

Berlin, July 1945

A flashlight revealed the stairs to the concrete underground air-raid shelter. Pools of stagnant water made the steps slippery. For an old man, uneasy on his feet, the descent was treacherous. Late in the afternoon on July 16, 1945, Britain's seventy-year-old wartime prime minister, Winston Churchill, picked his way with a gold-headed walking stick to the dark, dank bunker where Adolf Hitler had put a bullet into his right temple two and a half months before.

Word had spread quickly that Churchill was in Berlin. By the time his convoy reached the Reich Chancellery, the small British party had swelled to a jostling mob as war correspondents and numerous Russian officers and officials pressed forward to join Churchill's entourage. Anxious to witness the final scene of one of history's greatest dramas, they followed the prime minister, who wore a lightweight military uniform and visored cap, into the sacked remains of the Chancellery and, later, out to the garden where the entrance to the bunker was located.

In one of his most famous wartime broadcasts, Churchill had said: "We have but one aim and one single irrevocable purpose. We are resolved to destroy Hitler and every vestige of the Nazi regime. From this nothing will turn us. Nothing." Now, the reporters hoped for a curtain speech from this master of the spoken word as he inspected the tangible evidence of his triumph. He had been fighting his way here in one way or another for more than a decade, and a statement from him would provide a thrilling end to the story.

Churchill had been a lonely voice in the wilderness during most of the 1930s, when his warnings about Hitler had gone unheeded. In 1940, Britain was already at war when he was called to serve as prime minister. Against seemingly impossible odds, at a time when France had fallen and Hitler's armies had overrun the Continent, Churchill led Britain as it fought alone. While Nazi bombs rained on London and Hitler boasted that he had crushed the panic-stricken British in their holes, Churchill's flights of oratory rallied his countrymen and offered hope that their plight might yet be reversed. After the Russians and then the Americans joined the fight on Britain's side, Churchill battled the "bloodthirsty guttersnipe"—as he referred to Hitler—for an additional four years.

By the end of the war in Europe, Churchill had accomplished what many people had once believed he could never do. At home in Britain, even longtime detractors agreed that he had saved the country. His personal story was all the more remarkable because he had spent so much of his adult life in political disrepute. The road to the premiership had been long, "and every foot of it contested." Frustration, exclusion, and isolation had often been his lot before he became prime minister when he was sixty-five, an age that qualified him to draw an old age pension.

The man who visited Hitler's bunker had recast himself in just five years as one of history's titans. Had Churchill died before 1940, he might have been remembered as a prodigiously gifted failure. On this day, he was at the apex of his glory. Yet thus far, he had appeared oddly detached and distracted. His bulbous, bloodshot, light blue eyes surveyed the devastation at the Chancellery, and he quietly asked a few questions of the Russian soldier who served as his guide, but he made no public comment. Finally, Churchill left reporters outside the bunker entrance as he followed the Russian soldier into the blackness.

He slowly made his way down the first flight of stone steps

toward the chamber where Hitler's body had been discovered, slumped over a sofa beside the lifeless form of his bride, Eva Braun, her lips puckered and blue from poison. Churchill hesitated when the Russian told him that two additional water-soaked flights remained. As if it were no longer worth the effort, he abandoned the tour without having seen for himself the site of his mortal enemy's suicide. Churchill sent the others in his party, which included his youngest daughter, Mary, to view it without him. Then he turned and slowly began to make his way back up.

He climbed with difficulty. Five years of war had left Churchill in ravaged physical condition. In 1941, he had suffered a heart attack, the first of several episodes of heart trouble. He had repeatedly been stricken with pneumonia; on one occasion, in 1943, he had lain at the brink of death, at General Eisenhower's headquarters in Carthage. There had been moments during the war when Churchill was so exhausted that he could barely speak, walk, stand, or concentrate. In 1944 Clementine Churchill had told a friend resignedly, "I never think of after the war. You see, I think Winston will die when it's over." She knew better than anyone that her husband had put all he had into this war, and she was convinced it would take everything. She had watched as, whatever the state of Churchill's health, the phones kept ringing and the red boxes laden with official papers were rushed in. When illness sapped his energies and made it difficult to work, he pressed harder. Defying predictions that he would soon have to hand over to a younger, stronger man, he had fought on with the whole strength of his gargantuan will. Lady Violet Bonham Carter, a friend of four decades, feared what the "last pull up the hill" must have cost him. As victory drew near, Churchill was so physically weak that soldiers had to carry him upstairs in a chair after Cabinet meetings.

Churchill emerged from Hitler's bunker under his own power, but when at last he reached the top of the stairs and passed through the door of a concrete blockhouse into the daylight, his hulking frame appeared so shaky and depleted that a Russian soldier guarding the entrance reached out a hand to steady him. The Chancellery garden was a chaos of shattered glass, pieces of timber, tangled metal, and abandoned fire hoses. Craters from Russian shells pocked the ground. In one of those craters, Hitler and his wife had supposedly been buried after Nazi officers burned their corpses. The rusted cans for the gasoline still lay nearby. Russians pointed out the spot where the bodies had been incinerated. Churchill paused briefly before turning away in disgust.

He moved toward a battered chair that had been propped against a bullet-riddled wall. One of the Red Army men claimed that it had belonged to the Führer. The hinges of its back were broken and the rear legs had buckled. Churchill tested it first with one hand before sitting. Gingerly perched on the front edge of the seat, he mopped his forehead with a handkerchief in the withering heat as he chewed on a cigar. When at length his daughter and the others came out of the bunker, Churchill was visibly eager to leave.

In front of the Chancellery, he was met by a cheering crowd of sightseeing British sailors and Royal Marines. The street was a sea of devastation, every roof bombed out. The glassless windows of gutted buildings stared blindly. Despite the applause that greeted him, Churchill's mind was in turmoil; his heart ached with anxiety. This uncomfortable, even painful feeling of disconnection from the general rejoicing was not a new experience. For months, he had lived with the thought that in spite of what others might believe, the struggle was far from over. As the war came to an end, he saw that Soviet Russia, still ostensibly Britain's ally, was fast becoming as dangerous potentially as Hitler's Germany had been and that a third world war was already in the making. Worse, he knew that the Americans did not understand, indeed did not wish to be told, what was happening. As in a nightmare, the man who had warned in vain of the Nazi threat was again trying desperately to call attention to an emerging enemy.

In no mood to speak of victory, he acknowledged the cheers by mechanically raising his right arm and forming the familiar V-sign. Then he climbed into a waiting car for the return trip to Potsdam, outside Berlin, where he was due to face his Soviet and American counterparts. The talks, for which Churchill had been militating since May, were to have begun that afternoon, but the arrival of the Soviet leader, Joseph Stalin, had been inexplicably delayed. Potsdam was being billed as a victory conference, but Churchill privately regarded it as a chance to stage a "showdown" with Stalin about Soviet territorial ambitions.

The Big Three—Churchill, Stalin, and President Franklin Roosevelt—had last met at Yalta, in the Crimea, in February. But by the time he had assured the House of Commons that Stalin meant to keep his promise given at Yalta of free elections in the countries the Soviets had liberated from Germany, Churchill had already begun to be troubled by doubts. He worried that by trusting Stalin he might have made the same mistake that his predecessor as British prime minister, Neville Chamberlain, had committed with Hitler.

In March, Berlin became the focal point of Churchill's concerns when the Supreme Commander of the Allied Forces in Europe, Dwight Eisenhower, informed him that rather than try to take the Reich capital with American and British forces, he intended to let the Russians get there first. Churchill moved at once to persuade Eisenhower that he was about to make a calamitous error with far-reaching consequences. In separate messages to Eisenhower and Roosevelt, Churchill argued that, as the Soviet armies were about to enter Vienna and overrun Austria, should they also be permitted to take Berlin it would strengthen their conviction, already alarmingly in evidence, that they were chiefly responsible for Hitler's defeat and that the spoils were rightly theirs.

Churchill had no doubt that Stalin, whom Eisenhower had also informed of his intentions, well understood the symbolic and political significance of the Reich capital falling into Soviet hands. So when Stalin fulsomely complimented Eisenhower and assured him that the Soviet Union would send only second-rate forces to Berlin—which, he underscored, had "lost its former strategic importance"—Churchill's worst suspicions were confirmed. Churchill implored Eisenhower to pay particular attention to what was obviously a lie on Stalin's part. But Eisenhower flatly refused. Eisenhower had Roosevelt's backing when he declined even to attempt to race the Russians to Berlin. The Americans were not a little annoyed at Churchill's insinuations about their Soviet ally's postwar designs. Roosevelt had long insisted that Stalin was a man of good will and good faith who wanted nothing but security for his country, and that if the US gave him whatever he asked for, he could be counted on to work for a world of democracy and peace after the war and to refrain from annexing any territory. Meanwhile, on the very day—April 1, 1945—that Stalin had written to congratulate Eisenhower on his sound thinking, he ordered two of his top military chiefs to capture Berlin posthaste.

Driving through the ruins of the former Reich capital, Churchill sat in silence. His car passed endless rows of saluting Russian soldiers, directional signs printed in Cyrillic characters, and red-bordered posters bearing the sayings of Stalin translated into German above his signature etched in vivid red. It was as if Stalin had branded Berlin, fashioned it into his personal war trophy. The Russians all seemed "high as kites," convinced, as Churchill had predicted they would be, that they were principally responsible for having won the war. The shattered capital, where street after street had been reduced to rubble, and where the stench of putrefying corpses and broken sewer lines fouled the air, was an image of the collapse of the Nazi empire. Churchill grimly perceived something more. To his eye, the ubiquitous, overpowering Red Army presence augured the rise of Soviet power in postwar Europe.

Hardly had Berlin surrendered to the Russians, on May 2, 1945, when Churchill had hatched a plan. Experience had taught him that Stalin was not a man to be swayed by arguments based on abstract principles. "Force and facts" were his only realities. Brutal and unscrupulous, Stalin would do whatever he perceived to be in his own interest. The Americans had penetrated 120 miles deeper into Germany than originally planned. Churchill believed that gave him the leverage he needed to settle things peacefully with Stalin. The trick was to postpone pulling American troops back to the previously agreed-upon lines until Churchill was satisfied about both the temporary character of the Soviet occupation of Germany, and conditions in the countries liberated by the Red Army. Germany as a whole had yet to surrender when Churchill urgently contacted President Harry Truman, who had succeeded to office upon Roosevelt's death in April. He urged the former vice president that they invite Stalin to a heads of government meeting to take place as soon as possible and that in the interim they maintain their troops in existing positions in order to show the Soviets "how much we have to offer or withhold." From this point, everything depended on Truman. Roosevelt and Eisenhower had failed to listen to Churchill about Berlin. Could he hope for a better response from Truman, whom he had not yet met?

Churchill's telegram was sent on May 6. The following day, German generals signed the act of unconditional surrender of all German land, sea, and air forces in Europe. On May 8, Victory in Europe Day, Churchill announced in a radio broadcast that the German war was at an end; only Japan remained unsubdued. London exploded in a paroxysm of celebration. From first to last, the prime minister was at the center of the festivities. When he cried out to a vast crowd assembled beneath his balcony, "This is your victory," the people roared back, "No, it's yours!" A tender man easily moved to tears whether of joy or sorrow, Churchill made no secret of his pleasure in lapping up the affection and admiration, yet he remained oppressed by forebodings as he awaited Truman's answer. He ended the long, emotionally charged day by sharing his concerns with a friend, the newspaper proprietor Lord Camrose, who had come to dine at 10 Downing Street. While the noisy celebrations continued in the streets, Churchill spoke somberly and confidentially of what would happen if Truman turned him down and the troops were withdrawn. Stalin would control seven European capitals in addition to Berlin: Prague, Budapest, Warsaw, Belgrade, Bucharest, Sofia, and Vienna. As far as Churchill was concerned, that could not be permitted.

When Truman replied the next day that he preferred to wait for Stalin to propose a meeting, Churchill refused to take no for an answer. He wanted a leaders' meeting as soon as possible and suggested that he and Truman confer first in London in order to present a united front. The new president, who seemed to have inherited his predecessor's unrealistic view of Stalin, demurred on the grounds that he wished to avoid any impression of "ganging up." Churchill warned bluntly that the Soviets had drawn down an "iron curtain" upon their front and that the rest of humanity had no idea of what was going on behind it. He argued that surely it was vital to come to an understanding with the Soviet Union or at least see how things stood before American troops retired to the agreed zones of occupation.

Truman sent an emissary to London to convey what he did not wish to put in writing: Truman wanted to see Stalin first—without Churchill. The British prime minister could join them later. When Churchill waxed indignant, the emissary, Joseph Davies, went so far as to blame him for having provoked Stalin with his abiding hostility toward the Soviet Union. Truman's representative maintained that in fact it was Churchill's attitude that "placed not only the future, but possibly the immediate peace in real danger."

Furious at being told that he, not Stalin, was threatening the peace, Churchill reminded Truman that a shared love of freedom ought naturally to align their two nations against the Soviet Communists, who followed a different philosophy. With Churchill threatening to break publicly with Washington if Truman dared to see Stalin alone, the president appeared to back down. He agreed to three-power talks, but at the last minute he declined Churchill's pleas to postpone the retreat of the American army at least until after the conference.

When Churchill returned to his rose-pink, lakeside villa in Potsdam following the visit to Hitler's bunker, he learned that Stalin had arrived in Germany. The leaders were to meet the following day at 5 p.m. in the former palace of the German crown prince. Due to Truman's decision to withdraw his troops, however, Churchill was to face Stalin without the bargaining counter he had been hoping for. As he prepared to go into the talks, he was further wrong-footed by the fact that he had no idea of how much time he had to get what he wanted from Stalin. Prior to coming to Potsdam, Churchill—who had promised the British people a general election as soon as Hitler was defeated—had fought the first general election campaign in a decade. Polling day had been July 5, but three weeks had been allotted to permit the service vote to come in before the total vote was counted. On July 25, while the Potsdam Conference was still in progress, Churchill was due to fly back to Britain (briefly, he hoped) to learn his political fate.

The rapturous reception he had received in the course of his thousand-mile electoral tour strongly suggested to him that he would still be prime minister when the second round of talks began. Everywhere Churchill had traveled that spring, multitudes had come out to see and thank him for what he had done in the war. Standing nine and ten deep, enthusiasts had waved flags, sung patriotic songs, and cried out, "Good old Winnie!" Repeatedly they had closed in on the hero's open car and the progress of the motorcade had been slowed to a walking pace. Churchill had commented at the time that no one who had witnessed his reception could have any doubt about the outcome of the poll.

But what if there was an upset? What if poor Conservative showings in recent by-elections foretold a general swing to the left in British politics, as certain commentators were suggesting? Churchill had encountered some heckling, particularly in the last days of the campaign. Had these been isolated incidents, or did they reflect broad sentiment for change as Britons considered what they wanted their lives to be like after the war? Churchill had brought his Labour opponent, Clement Attlee, to Potsdam to make it clear that all of Britain was being represented at the conference table, as well as to ensure that there would be no break in Attlee's knowledge of affairs. But by his very presence, Attlee was also a reminder that when Churchill went home for the election results, they might prevent him from completing what he had begun with Stalin.

Attlee, who had been deputy prime minister in Churchill's wartime coalition government, attended a small luncheon at the prime minister's villa on the afternoon of July 17. Also present was US Secretary of War Henry Stimson. Churchill lingered at the table bantering with his guests until half an hour before he was scheduled to see Stalin and Truman. With Attlee, he joked about the British general election as if there had not been a harsh word between them during the often exceptionally bitter campaign; and with Stimson, he exchanged light-hearted personal reminiscences about their younger days. In the leisurely course of the meal, Stimson gave no sign that he had anything urgent to convey to his host or that he desired to speak to him privately. Only when Churchill was showing Stimson out did the latter inform him that the world's first atomic bomb had been successfully tested the day before in New Mexico. Though Britain had been a partner in the secret project to develop the new weapon, deadlier than any yet in existence, Churchill had not known in advance the date of the test. If the bomb worked, it promised to end the Pacific war quickly and to spare a great many Allied soldiers' lives. In his present predicament, Churchill also sensed that it might be capable of something more. Nothing could be certain for there were no details as yet, but here potentially was the card he needed to persuade Stalin to come to terms in Europe.

Until Churchill had additional information, he was intent that this first bit of news from New Mexico be withheld from Stalin. Truman had already thrown away one bargaining counter, and Churchill wanted to be sure that it did not happen again. Stimson's insistence that the Soviets really ought to be notified of the test left Churchill to go off to his meeting both excited about the new development and worried that the Americans might be about to squander another opportunity. Churchill, who had met with Truman the previous day, knew that the president had been lunching with Stalin that afternoon. Churchill had no idea whether Truman, who had had the information since the previous evening, might already have told Stalin about the bomb or even agreed to share its secrets with his Soviet ally, as Roosevelt had once been inclined to do. The last-minute timing of Stimson's disclosure made it impossible for Churchill to talk to the president beforehand.

At the first plenary session at the Cecilienhof Palace—in whose courtyard the Russians had planted masses of geraniums in the form of a 24-foot-wide red star—Churchill played for time. For weeks, he had been pressing Truman to agree to the earliest possible meeting; now, he meant to go as slowly as possible. Previously, he had been insistent that the most contentious issues be addressed without delay; now, he told his fellow leaders, "We will feel our way up to them." Truman was in a difficult, almost impossible position at Potsdam. In the days before the conference, he had been nervous about facing giants like Churchill and Stalin, and with good reason. Though a dying man, Roosevelt had done nothing to prepare his successor for the presidency. Roosevelt never spoke to Truman confidentially of the war, of foreign affairs, or of what he had in mind for the postwar world. When Truman suddenly landed in the Oval Office after only three months as vice president, he had to struggle to catch up. Devoid of foreign policy experience, he pored over memoranda, briefs, and correspondence on international affairs. He conferred with presidential advisers, who, to his perplexity, offered conflicting advice. On the voyage to Europe, he absorbed as much information as he could from various coaches. At his first meeting with Stalin and Churchill together, he read aloud from prepared statements and thereafter was careful to stick to positions that had been well worked out in advance.

Churchill's newly unhurried pace irked Truman, who protested, "I don't just want to discuss. I want to decide."

Churchill came back impishly, "You want something in the bag each day."

Looking ahead to subsequent plenary sessions, Truman went on, "I should like to meet at four o'clock instead of five."

"I will obey your orders," Churchill replied.

Clearly amused, Stalin interjected, "If you are in such an obedient mood today, Mr. Prime Minister, I should like to know whether you will share with us the German fleet."

When the meeting concluded, Stalin invited the others to the banquet room, the length of which was filled by a table loaded with caviar, cold meat, turkeys, partridges, and salads, as well as vodkas and wines "of all hues." Truman stayed less than ten minutes. Churchill, though he had previously complained of indigestion, happily ate and drank with Stalin. Pointing out that they had much to talk about privately, Stalin asked Churchill to have dinner with him the following evening. It promised to be a late night. In contrast to Truman, who preferred to go to bed early, Churchill and Stalin were both night owls. On the present occasion, Stalin remarked to Churchill that he had grown so accustomed to working late during the war that even though the necessity had passed, he could never get to sleep before 4 a.m.

By the time Churchill dined with Stalin on Wednesday, July 18, he had had a chance to talk to Truman, who assured him that he had not informed the Soviet leader about the bomb. Churchill had concluded overnight that, once they were certain of the test results, it would actually be a very good idea were Stalin to be made aware that the Western Allies had a singular new weapon. What continued to worry him, however, was the possibility that Truman would agree to share technical information with Moscow. Truman said he would not, but Churchill remained uneasy. Western possession of the bomb would be of little use in the negotiations if Stalin could count on being able to build one as well.

When Churchill arrived at half past eight, Stalin's two-story stone villa was surrounded by machine-gun-toting thugs. A phalanx of seven NKVD—secret police—regiments and nine hundred bodyguards had accompanied the dictator to Potsdam. Stalin had lived a life of violence, fighting off rivals and would-be usurpers with murder and bloodshed, and he was perpetually fearful of being treated in kind. In the present setting, the savage vengeance the Red Army had wreaked on the Germans provided an additional motive for an assassination attempt. Still, once through the numerous layers of security, Churchill was welcomed with friendly informality. Stalin had no other guests. Only the leaders' interpreters, Birse and Pavlov, were scheduled to join them at dinner.

Though the surviving members of the original Big Three had become antagonists, they were bonded by a sense of themselves as men apart, the last of a superior breed that had included Roosevelt. As Churchill later said, together they had had the world at their feet and commanded many millions of men on land and sea. Churchill's personal history with Stalin had begun with their written exchanges in 1941 after Germany's surprise attack on Russia. When they first met, in Moscow in 1942, Stalin's insults had nearly caused Churchill to break off their talks and fly home at once. The British ambassador, Archibald Clark Kerr, had worked hard to assuage Churchill's fury. Kerr urged him to reflect on the relative unimportance of his own wounded feelings weighed against the many young lives that would be sacrificed if he did not swallow his pride and return to the talks. After that, Churchill resolved that in the interest of advancing their shared objective of defeating Hitler, he would do what it took to build a relationship with Stalin, even somehow find a way to "like" him. In subsequent meetings—at Tehran, again in Moscow, and at Yalta—Churchill and Stalin had developed a deep fascination with and respect for each other.

Though Churchill had no idea that Stalin's late arrival at Potsdam had been due to a minor heart attack, he thought his host tonight looked ill and "physically rather oppressed." Stalin's once black hair was a grizzled mop, as were his shaggy mustache and eyebrows. His narrow, evasive eyes were yellow, his teeth stained and broken. Short and stocky, he had a withered left arm that he held woodenly in his right hand, palms up. War had aged the sixty-seven-year-old Generalissimo prematurely. He had worked too hard, slept too little, and not had a vacation in years. Despite his physical condition, he had made no more effort to moderate his appetites than Churchill had. In the course of the dinner, which extended to the next morning, there was a prodigious amount of smoking, eating, and drinking by both men.

When Churchill and Stalin last saw Roosevelt, at Yalta, the

president had been a cadaverous figure with waxen cheeks and trembling hands. At times he had been lucid, but at other moments he sat with his mouth open, staring ahead, unable to follow the discussion. Roosevelt had clung to office in the belief that he was indispensable. Five months later, the conversation between Stalin and Churchill naturally turned to the death of great men and the problems of succession. A discussion of how Roosevelt's successor was doing under the circumstances prompted Churchill to ask if Stalin had given any thought to what would happen at the Kremlin after he died.

Churchill had designated his own political heir. In 1942 he had written to King George that in the event of his death he wanted Anthony Eden to carry on as prime minister because he possessed the "resolution, experience and capacity" the times demanded. Tall, slim, graceful, debonair, the forty-eight-year-old foreign secretary accompanied Churchill to Potsdam. Eden had a furrowed but handsome face with penetrating pale blue eyes and a carefully trimmed gray mustache, and he spoke in a mellifluous baritone. Stalin too had brought an heir apparent to the Potsdam Conference: Eden's opposite number, the fifty-five-year-old Soviet foreign minister Vyacheslav Molotov. Recently, Stalin had anointed Molotov with the words, "Let Vyacheslav go to work now. He is younger." Small and chunky, Molotov spoke with a slight stammer and had an impassive, "lard-white" face. When he was upset, though his countenance remained stony, a telltale lump in his forehead swelled and throbbed alarmingly.

Stalin insisted to Churchill that he had arranged everything. He claimed to have groomed good men and thereby to have guaranteed the continuity of Soviet policy for thirty years. He made it all sound so sensible, but then Stalin was adept at portraying himself to foreigners as utterly reasonable and rational. In fact, in the words of the American diplomat George Kennan, he was "a man of absolutely diseased suspiciousness." Stalin had a history of exterminating not only his opponents but also those whose character suggested that they might oppose him later. He sniffed plots and cabals everywhere, and never more so than after the war. In 1945, the leader widely venerated as his nation's savior was at the pinnacle of his power. At the same time, his own physical decay left him feeling vulnerable to the machinations of the ambitious younger men who formed his circle. As the fawning Molotov and other contenders for the postwar Soviet leadership well knew, Stalin was capable of ordering their arrest or execution at any time, "no questions asked." (As a precaution, Molotov slept with a loaded revolver under his pillow and never permitted his sheets or blankets to be tucked in lest he have to leap out of bed and defend himself in the middle of the night.) Stalin spoke matter-of-factly of retiring on a pension in two or three years, but he was no more inclined to leave office willingly any time soon than Churchill was.

This evening, when Churchill voiced anxiety about the British election, Stalin expressed confidence that the Conservative leader had nothing to worry about. The concept of free elections meant nothing to Stalin. Churchill was in power; surely he had arranged to stay there. The only real mystery as far as the dictator was concerned was why Churchill would go to the trouble of flying home for the result. Characteristically, Stalin suspected a ploy on Churchill's part.

Stalin was right to sense that Churchill was up to something, though at this point the latter's calculations had nothing to do with the British election. Churchill asked whether there would be free elections in the territories under Soviet control, and he raised concerns that the Red Army was preparing to surge westward across Europe. When Stalin sought to reassure him on every count, Churchill took care not to provoke an angry confrontation by too directly challenging anything the Generalissimo said. Churchill was biding his time until he knew what kind of hand he had to play with Stalin.

Exactly a week remained before Churchill was due to fly to London. Every day that passed without news from New Mexico with precise details of the bomb test was an agony to him. There were plenary sessions on Thursday and Friday, but still no additional information came in. At half past four on Saturday, Churchill was about to leave for the Cecilienhof when Stimson arrived with the full report. This was the document the prime minister had been waiting for since Tuesday; everything depended on its contents. But no sooner had he begun to read than an aide reminded him that if he did not go now, he would be late for his 5 p.m. meeting. Following the day's talks, Stalin was due to host a party for all of the conference's participants, and Churchill naturally was expected to attend.

As the report was of the highest secrecy, Stimson had shown the single copy personally to each individual on his list, beginning with Truman. There was no question of his leaving the document for Churchill to study later, so it was agreed that he would return to the prime minister's villa in the morning. Churchill reluctantly handed the report back to Stimson. Impatient to resume reading, he found the evening that followed interminable.

It was not until 11 a.m. on Sunday that Stimson reappeared and Churchill at last had a chance to study the report in full. The details of the atomic bomb gripped him: the lightning effect equal to that of seven suns at midday; the vast ball of flame which mushroomed to a height of more than ten thousand feet; the cloud which shot upward with immense power, reaching the substratosphere in about five minutes; the complete devastation that had been wrought within a one-mile radius. Immediately, Churchill saw that this was the card he had been hoping for. The bomb completely altered the balance of power with the Soviets. Stalin's vast armies were negligible compared to it. Truman no longer had to worry about Stalin's willingness to fight the Japanese, and Churchill hoped that that would translate into real support for some tough bargaining to get a viable settlement in Europe. He rushed over to see Truman, both to discuss a speedy end to the war in the Pacific and to confirm that the Americans were not intending to share the bomb's secrets.

Churchill spoke excitedly of the bomb to his physician, Lord Moran, the following morning. He swore the doctor to secrecy and assured him that it had come just in time to save the world. Again at lunch, he laid out the new situation to Field Marshal Sir Alan Brooke, General Hastings "Pug" Ismay, Eden, and other key members of the British delegation. Referring to the sudden shift in the diplomatic equilibrium, Churchill thrust out his chin and scowled. He spoke of threatening to blot out Russian cities if the Communists refused to behave decently in Europe. But for all his talk of bullying Stalin with the bomb, Churchill's aim was not to start another war. As he had told Eden early on, he believed that the right bargaining counter might make it possible to secure a "peaceful agreement." He calculated that Stalin did not want war any more than he did, only the fruits of war, which the Soviets felt they had earned by their signal contribution to the defeat of Hitler. If Stalin could not be persuaded to settle, it might be best, as Churchill had previously told Truman, to at least know where they stood with him—and to know it sooner rather than later.

Churchill's optimism about what he would be able to achieve with both his fellow leaders provoked intensely skeptical reactions from British colleagues. There was sentiment in the British camp that Truman (who controlled the bomb, after all) would never provide the backing Churchill needed, that Stalin would simply shrug off any real or implied threat, and that the details in the report from New Mexico might yet prove to have been exaggerated. Still, Churchill had found reason to hope, and to him that was all that mattered. On Monday night he called Lord Beaverbrook, who had had a hand in shaping the Conservatives' electoral strategy, for the most up-to-date predictions. Churchill had come to Potsdam empty-handed; now that he had what he believed was the basis of a real negotiation, nothing must be allowed to interfere. Beaverbrook told his friend that the Conservatives were expected to win, though perhaps by a smaller majority than first predicted.

Having proposed at the outset that the leaders take their time moving toward the most difficult questions, Churchill was ready to step up the pace and intensity of the talks. But the moment was still not right for what he saw as the climactic confrontation about Soviet intentions in postwar Europe. That, he believed, must wait until the British election results were known and the people had affirmed their confidence in him. Fresh from having submitted himself to their judgment, he would be in an optimal position to demand free elections in the territories liberated by the Russians.

He managed to put off the sharpest exchanges of the conference until Tuesday, July 24, the eve of his departure. Speaking of reports from Romania and Bulgaria, he charged that an "iron curtain" had descended in those countries. Until this point in the talks Stalin had been inclined to speak in a low, controlled tone of voice, but Churchill had succeeded in arousing his ire, and he shot back, "Fairy tales!" A fierce dispute about the veracity of Churchill's claims followed. There was a good deal of pique and perspiration on the Soviet side of the large round table, which was covered with a dark red felt cloth and arrayed with offerings of pungent Russian cigarettes. Both Molotov and Eden grew indignant on behalf of their respective masters. Eventually Stalin declared that his and Churchill's views were so far apart that the discussion ought to be broken off—for now.

After everyone rose, Churchill watched anxiously as Truman walked over to Stalin. Churchill and Truman had previously agreed that at the close of that day's session the President would tell Stalin about the bomb and the plan to use it on the Japanese. (In fact, Stalin's spies had already notified him of the successful test blast, but neither Churchill nor Truman knew that.) There was high tension as Churchill looked on from a distance of about five yards. He longed to see Stalin's reaction, but he was also watching Truman. What would the president do if pressed for technical information? Would he agree to a meeting of American and Soviet experts? Truman had said in advance that he would not, but Churchill was aware that there had been no firm promise and that Truman did not yet perceive the Soviet threat as he did. Both participants in the silent scene were acting: in an effort to seem as casual as possible, Truman had left his own interpreter behind and depended on Stalin's man to translate his remarks, while Stalin made a point of appearing by turns genial and nonchalant.

Later, as the leaders waited for their cars, Churchill found himself beside Truman. He inquired how the conversation had gone. Truman reported that Stalin had not so much as asked a question. Stalin had said only that he was glad to hear the news and that he hoped they would make good use of the new weapon against Japan.

In any event, the information had been conveyed, and every element was finally in place for the dramatic confrontation Churchill expected would occur after a forty-eight-hour intermission. He was in buoyant spirits when he dined with Lord Mountbatten, the Supreme Commander in South-East Asia. Churchill again had much to say about the bomb and his plans for the future, though Mountbatten wondered whether the prime minister might not be assuming too much about the election outcome. That Churchill may have had deep doubts of his own is suggested by a disturbing dream he had that night. Six nights after he and Stalin had talked of death and succession, he dreamed that he too had died. He could see his corpse laid out beneath a sheet in an empty room. The face and body were draped, but the feet that stuck out were recognizably his own. On Wednesday morning, as he prepared to attend a final brief meeting with Stalin and Truman, he feared the dream meant that he was finished.

To all outward appearances his confidence had been restored by the time of the ninth plenary session. At a quarter past twelve, when Truman adjourned the meeting until 5 p.m. on Friday, Churchill added crisply that he hoped to be back. His mood on the flight home with his daughter was one of certainty that he would soon return to complete what he had begun. In London, Churchill went to Buckingham Palace to report to the King on the talks thus far and on the changes in the international situation that the bomb had wrought. Before Churchill retired for the evening at the Annexe facing St. James's Park, he was pleased by the political gossip that even Labour headquarters was predicting a Conservative majority. Fittingly, he intended to monitor the figures from the Map Room, where once he had tracked the unfolding of the Allied victory over Hitler. Family members and close friends had been invited to sit with him as numbers streamed in throughout the day on July 26.

Churchill went to bed on Wednesday night convinced that those numbers would favor the Conservatives. (Rather than vote directly for a prime minister, voters selected party members in their individual constituencies.) Sometime before daybreak, he awakened suddenly with a stabbing pain that told him the election was lost and the power to shape the future would be denied him. In anguish, he rolled over and slept until nine. About an hour later, Churchill was in his bath when he learned from an aide that his premonition of disaster was being amply confirmed by the early poll figures.

After he had dressed in a blue one-piece zip-up siren suit, he went to his Map Room. Over the next few hours, Churchill, surrounded by charts of constituencies and of the most recent state of the election, reacted to the news of each Labour gain by silently, stoically nodding his head. He complained only of the heat and the want of air. The Conservatives were out. Churchill had been returned in his constituency, Woodford. But overall there had been a Labour landslide, and Britain was to have a new prime minister.

Mrs. Churchill, tall, silvery-haired, with proud posture and a profile said to resemble a ship's prow, suggested to her husband that the outcome of the election might prove to be "a blessing in disguise."

He replied, "At the moment it seems quite effectively disguised."

It was inconceivable that he had been cut off altogether from Potsdam. Initially, he insisted he would wait to take his dismissal from the House of Commons, as he was entitled to do. Then in his pride he declared that nothing would induce him to go back to Potsdam, though he was not yet ready to resign immediately either.

He spoke vaguely of stepping aside on Monday, though that would mean asking Stalin and Truman to wait in Germany until he made up his mind. Finally, Churchill accepted that under the circumstances he really had no choice but to resign at once—and let the talks go on without him. All of his great plans, everything he had so carefully set up, must remain unrealized.

Some twenty-four hours after he had raced to Buckingham Palace to speak to the King of his hopes, Churchill returned to tender his resignation. The King offered him the Order of the Garter but Churchill declined the high honor in the belief it would be wrong to accept it after what he saw as a public rebuff of his leadership. He drafted a statement to be read aloud to the nation on the BBC at 9 p.m. He stated that as a consequence of "the decision of the British people," he had laid down the charge which had been placed upon him in darker times.

Thursday was devoted to immediate concerns. The next morning, he awakened to the realization of what the people's decision meant for him personally. In years past, Churchill had been known to declare that in war one can only be killed once, but in politics many times. Politicians, he once wrote, expect to fall and hope to rise again. In the face of staggering rejection, numerous setbacks and many apparent dead ends, obstinacy had kept Churchill pounding on when fainter spirits might have given up. "No" was an answer he had repeatedly refused to accept. "Unsquashable resilience" had long been among his defining characteristics. He had justly been said to have more lives than a cat and to have survived as many arrows as legend planted in the flesh of Saint Sebastian.

This time everything seemed different to him, and it was his age that made it so. By most estimates the magnitude of the Labour victory, a majority of 146 seats in the House of Commons, suggested that the Conservatives could not hope to return to power for at least a decade. Some people went so far as to say that Labour was in for a generation. For Churchill, as for his party, there was no avoiding the likelihood that by the time he had another chance at the premiership—if he ever did—he would be at least eighty.

Throughout the day, as he said goodbye to some of the people who had worked most closely with him during the war, he seemed absorbed by the idea that a comeback was impossible. The previous night he had briefly thought the new Labour government might be turned out soon enough, but the final numbers left no such hope. After a farewell meeting with his Cabinet, he lingered privately to talk to Eden. Churchill expressed confidence that the Conservative heir apparent would surely sit in the Cabinet Room again.

"You will," Churchill said with more than a dash of bitterness, "but I shall not."

Churchill once observed that a man's only real necessities in life are food and a philosophic temperament. All day Saturday at Chequers, the official country residence of British prime ministers which he had yet to vacate, he appeared remarkably cheerful and controlled, but after dinner and a film screening, his mood darkened noticeably. Attlee had returned to Potsdam a day later than scheduled and the talks resumed that very night at ten. The new Big Three worked at the conference table in the Cecilienhof until after midnight.

Cut off from all that forever, it seemed, Churchill sat into the night listening to Gilbert and Sullivan and other recordings on his gramophone.

11

FACE FACTS AND RETIRE

London, 1945

During the war, Churchill had always slept with the key to the Cabinet boxes. He kept it on his watch chain. Suddenly, the key was gone. There were no more boxes, no constantly ringing phones, no telegrams requiring his immediate attention. From the first, he found it impossible to adjust to a new tempo of life. It was as if his heart was still pounding at the maximum rate, though his body had been forced to a standstill. Previously, Clementine Churchill had doubted that he could survive the intense demands on his time and energy. The question now was whether he could live without them.

Churchill returned to London on Monday, July 30. While Clementine supervised the removal of their possessions from 10 Downing Street, the couple took up temporary residence in a sixth-floor penthouse at Claridge's hotel in Mayfair. Long prone to bouts of depression, Churchill worried about sleeping near a balcony at a time when he was haunted by what he described as "desperate thoughts."

A sense of incompleteness gnawed at him. He could hardly believe that other hands had taken over with Stalin when it was he, Churchill, who had established a relationship with the Soviet leader, understood him, and saw what needed to be done. Not for the first time in his life, Churchill despaired that his peculiar powers and gifts were being wasted.

The shift in his political fortunes had left him feeling hurt, humiliated, and confused. Because he had cast the election as a referendum on his conduct of the war, he was tortured by the idea that the outcome had called his record into question. Though he was no stranger to rejection, he had never developed the thick skin that is so useful to a politician. No analysis of the Labour vote prevented him from interpreting the numbers as a personal disgrace. No references to an overall leftward swing in British politics, to the public's desire for a better material life after the war, or to festering public resentment of previous Conservative leaders' failure to prepare adequately for the struggle against Hitler were capable of lessening the blow to Churchill's ego.

For all that, he was able to make light of his defeat, as when he sang an old music hall song to the doorman at Claridge's:

> I've been to the North Pole
> I've been to the South Pole
> The East Pole, the West Pole
> And every kind of pole
> The barber's pole
> The greasy pole
> And now I'm fairly up the pole
> Since I got the sack
> From the Hotel Metropole.

Sometimes, others sang to him. On August 1, six days after the Conservative rout, the new Labour-dominated Parliament assembled to elect the Speaker of the House of Commons. The chamber was packed and tensions ran high as, flushed with victory, the "new boys" on the Labour benches taunted and jeered at the Conservatives sitting across from them. At a quarter to one, when everyone else was in place, Churchill made his much anticipated entrance. With chin up, he sauntered to his seat on the opposition front bench. Conservatives greeted him with raucous cheers. One Tory began to sing "For He's a Jolly Good Fellow," and others heartily joined in.

Labour countered with their party anthem, "The Red Flag." In boisterous spirits, a Labour member dashed to the front to pretend to conduct what a witness likened to "the chorus of birds and animals sometimes to be heard in a Disney film." Some of the newcomers did not really know the words and struggled to improvise. Meanwhile, whenever the Labour majority threatened to drown the Conservatives out, the latter raised their voices.

At Chequers the previous weekend, Churchill had made it clear that he hoped to go on leading the opposition and the party so long as Conservatives wished him to and so long as his strength held. The cheers from the Tory benches suggested that he had his party's unanimous support. The British people might have rejected him, but the emotional reception accorded him by fellow Conservatives left him in no doubt that they at least wanted him as their leader.

Sadly, as he had done at the time of his electoral tour, he misconstrued the meaning of those cheers. Now, as then, the fact that people were grateful to their wartime prime minister did not necessarily mean that they wanted him to remain in power during what promised to be very different circumstances.

Hardly had Churchill left Westminster after his debut as opposition leader when a trio of Conservative heavyweights met to find a way to maneuver him out of the job. The driving force in the effort was Robert Cecil, Viscount Cranborne, leader of the opposition peers. The fifty-one-year-old heir to the 4th Marquess of Salisbury was a grandson of the prime minister Lord Salisbury and a cousin of Churchill's wife. His politically prominent family had been advising monarchs since the sixteenth century and one ancestor had been instrumental in the accession to the English throne of James VI of Scotland. In the Conservative Party as well, the Cecils (pronounced to rhyme with "whistles") were noted as kingmakers and power-brokers. They had had a long and complicated relationship with the Churchill family. "Your family has always hated my family," Winston was known to grumble, at which Cranborne would "laugh uproariously" in response. In 1886, the prime minister Lord Salisbury had been responsible (at least as Winston saw it) for the political ruin of Winston's father, Lord Randolph Churchill, the second surviving son of the 7th Duke of Marlborough. In the 1930s, Cranborne himself had done much to obstruct Winston's career despite the fact that both men opposed the policy of appeasement. Nevertheless, Winston had long found it impossible to decide whether he admired the Cecils or resented them on his late father's behalf. Salisbury had been kind to Winston when he was in his mid-twenties, and Winston later dedicated a book to him. Winston had enjoyed friendships with Cranborne's uncle, Hugh Cecil, who served as best man at Winston's wedding, as well as with, however improbably, Cranborne himself. He adored Cranborne's wife, Betty, whose acerbic conversation he prized, and through the years he and Cranborne had managed rigorously to keep their political differences in one compartment and their private friendship in another.

Known familiarly by his boyhood nickname, Bobbety, Cranborne was tall and gaunt, with a long nose and protruding teeth. A speech impediment caused his r's to sound like w's and he spoke at the breakneck speed characteristic of his family, who were famous for their ability to utter more words in a minute than most people can in five. He was an ugly man, slightly bent and often shabbily dressed, whose great personal charm caused many women to find him immensely attractive. He was an invalid and hypochondriac, whose frail frame housed a will of iron. And he was a political powerhouse, who, like other Cecils before him, preferred to operate behind the scenes, often so subtly that it was difficult to perceive his hand in events. Invisibility appealed to him because he prided himself on basing his actions not on the dictates of personal ambition, but on duty and principle. Since he ostensibly wanted nothing for himself, he had a reputation in party circles for "objectivity" that gave his pronouncements particular weight. Past disagreements notwithstanding, Cranborne warmly acknowledged the greatness of what Churchill had done in the war. He marveled, as he later said, that in 1940 Churchill "did not talk of facing the realities: he created the realities." Now that victory had been secured, however,

Cranborne maintained that Churchill ought to "face facts and retire" without delay. Had not Churchill himself taken a similar view of Cranborne's grandfather in the twilight of his career?

During the war, the Tory party had been allowed to disintegrate on almost every conceivable level. Churchill was not a party man and never had been, and from the time he became the party's leader in 1940 he had shown no interest in overseeing its affairs. As a consequence, by 1945 there was no management, no organization, and no program. Lacking specific policies, Tories had fought the general election on the aura of the Churchill name and record. In the wake of overwhelming defeat, there was broad agreement that Conservatism needed to be drastically rethought. In Cranborne's view, the effort needed to begin immediately under younger leadership. Anthony Eden had long been his "horse" in the political race. From the outset, Cranborne's career in politics had been closely tied to Eden's. Cranborne started out as Eden's parliamentary private secretary, and Eden and he grabbed headlines together in 1938 when they resigned as foreign secretary and under secretary of state for foreign affairs respectively, in protest at the Chamberlain government's appeasement of Mussolini.

The description of Cranborne as Eden's under secretary belied a more intricate relationship. Cranborne was Eden's most powerful political supporter. Cranborne did not himself aspire to the premiership; influence was what he was after, and he viewed an Eden government as the best way to secure it. As a politician, Eden benefited from the prestige of a connection to the house of Cecil, as well as from Cranborne's superior intellect and cunning. Cranborne was also the nervier by far. He often quietly but insistently pressed Eden to act as Eden almost certainly would never have dared on his own. Eden, whose theatrical good looks and sartorial elegance contrasted sharply with Cranborne's less photogenic appearance, attended the meeting on August 1, 1945, to discuss Churchill's future.

Also present was Britain's ambassador to the US, Lord Halifax, who had returned from Washington on leave the previous day.

When Chamberlain resigned in 1940, Halifax had been his choice to replace him as prime minister, as well as that of King George VI and much of the Conservative Party, but Halifax had declined in favor of Churchill. (Halifax calculated that the public clamor for Churchill was so great that any other appointment would inevitably be overshadowed by his looming presence in the government. Halifax took consolation in the judgment that Churchill's character flaws made it likely that his tenure would be brief.) When Churchill became prime minister, he returned the favor by shipping Halifax off to Washington lest he reemerge as a political rival.

For Churchill, who could be as passionate about fighting off real and potential rivals for power as he was about fighting the war, there had been an additional advantage to replacing Halifax as foreign secretary with Anthony Eden. As the Conservative chief whip James Stuart later observed, Churchill "knew he could bully Anthony . . . but not Halifax." By exiling him to the US, Churchill lowered the curtain on Halifax's political career. Five years later, Halifax was one of those who believed the time had come for Churchill to bow out, and by his reckoning, Churchill was fortunately not one of those individuals whose sole interest in life is his work. There were many activities that afforded him much pleasure, but that he had had little time to pursue during his premiership. Among other things he was an author and painter, and Halifax believed he might actually welcome a chance to be free of the burdens of leadership and retire of his own accord.

Eden thought he knew Churchill's mood better. Cranborne, as well, was far from optimistic that Churchill would willingly step down, and his strategy was to ease him out of power. Churchill had been asked to go to New Zealand to be honored for his war service, and Cranborne was determined that he accept that invitation, as well as many similar ones that were sure to follow from around the world. While Churchill was abroad, Eden would run the party in his stead. Halifax was set to see Churchill at 5 p.m. that day, and Eden was to dine with him after that. Cranborne urged both visitors to press Churchill to go to New Zealand, and generally to entice him with the joys of retirement. Halifax readily agreed, but Eden hesitated.

This was partly a matter of propriety on Eden's part, partly a matter of self-preservation. He flinched at the unseemliness of trying to push Churchill aside in open pursuit of his own interests. He did not wish to appear vulgarly ambitious. Was that not among the very qualities in Churchill that had long repelled him and many others? At the same time, Eden longed to lead the opposition and he did not want to do or say anything to provoke Churchill to turn against him at this late date and to name another successor. During the war, Churchill had been known to taunt him with the names of other "possibles"—Oliver Lyttelton, John Anderson, Harold Macmillan. By Eden's lights, the wait to succeed Churchill had been long and excruciating, and he did not want to jeopardize his position before the handover actually took place. But Cranborne was insistent, and as had often been the case between the two friends, Eden reluctantly gave in to the stronger will.

As it happened, August 1 was also the closing day of the Potsdam Conference. The plenary sessions were set to wind up that night, and Halifax was scheduled to be present at the King's meeting with Truman in England the following day. The talks had failed to produce anything like the settlement Churchill had been chasing. When the newly configured British delegation returned to Potsdam, Stalin had viewed Attlee warily and had insisted on grilling him and Ernest Bevin, the new foreign secretary, about Churchill's fate. Long preoccupied with averting his own removal from office, Stalin was palpably shaken by the news from London. If Churchill was dispensable, presumably the same was true of Stalin. He did not like the change. He would not have it. For two days he failed to show up at the conference table. Ostensibly he was ailing but Truman suspected the real reason was that he was upset about Churchill. When Stalin did reappear he seemed to have lost interest in the proceedings.

Under the circumstances, Potsdam and all that it signified to Churchill cast a shadow over Halifax's visit to Claridge's. Rather than listen to his guest's paean to memoir-writing and other activities to be looked forward to in retired life, Churchill preferred to mourn the loss of power and efficacy. In conversation, he dwelled on the fact that only a week had passed since he had been at Potsdam. He found that impossible to believe. How could everything have changed so quickly?

By the time he saw Eden, he had begun to dwell on the mistakes that might have led to his defeat. It was widely believed that by attacking Labour too violently Churchill had diminished his hard-won status as a national hero who stood above the political fray. On the present occasion, he lamented that if only Eden had not been ill with a duodenal ulcer during the campaign he would have had the advantage of his heir's advice and avoided that perhaps fatal error. In making such a claim, Churchill was flattering Eden in the conviction—which had helped to sustain the number two man during the war—that he served as a "restraining hand" on Churchill's often monstrously poor judgment. (As Pug Ismay once put it, "Some men need drink. Others need drugs. Anthony needs flattery.") Over the course of the evening, Churchill clutched Eden to his bosom, insisted the younger man was his "alter ego," and otherwise strove to convey how much he valued and depended on him. But the love fest was short-lived. When Eden dared to suggest that the party vice-chairmanship be awarded to a close friend of his own, a man associated in people's minds with Eden's interests, Churchill exploded. The charm, the flattery, the unctuous affection—all dissolved in an instant. Furious at being pushed, Churchill made it clear that he had his own candidate for the post.

After Eden left at midnight, Churchill swallowed a sleeping pill and went to bed. Since Saturday when the Potsdam talks had resumed without him, he had found that even after taking a "red" he was unable to sleep through until morning. For the fifth night in a row, he shot awake at 4 a.m., his thoughts racing uncontrollably, and he required a second barbiturate pill to sleep.

In the days that followed, Churchill in his misery was of various minds about how to proceed. He tested some of the suggestions others had made, but none appeared to satisfy him. He spoke of his war memoirs, but despaired of the taxes he would be required to pay on the earnings. He said he might go abroad indefinitely, but complained that he had no appetite for travel. He talked of honors and invitations, but added that he was in no mood to accept them. By turns he vowed to fight to retain control of the Conservative Party and admitted that he did not know how much longer he wished to lead. He groused that it would have been better had he died during the war but also insisted that he was not ready to die.

On August 6, the first atomic bomb fell on Hiroshima. A second bomb targeted Nagasaki three days later. The new weapon, which reduced innumerable victims to tiny black bundles of smoking char, proved to be everything described in the report Churchill had seen at Potsdam—and more. The news jolted him out of his lassitude and self-pity, for it vividly suggested what a world war fought with atomic weapons might be like. The prospect made it more urgent than ever to get a proper postwar settlement in Europe, but political defeat had left Churchill helpless to do anything except try to warn the world in time.

Japan surrendered unconditionally on August 14. That evening, Churchill dined with Eden and other Conservative colleagues in a private room at Claridge's. After the meal, he and his guests listened to Attlee's midnight broadcast to announce the end of the war. Inviting people to relax and enjoy themselves in the knowledge of a job well done, Churchill's successor decreed two days of victory holidays.

V-J Day, August 15, coincided with the state opening of Parliament. An estimated one million celebrants lined the route from Buckingham Palace as the King and Queen rode to Westminster in an open red-and-gold coach drawn by the Windsor grays. Elsewhere, in a scene reminiscent of his thousand-mile electoral tour, Churchill, traveling in an open car, was mobbed by well-wishers. Shouting "Churchill forever" and "We want Churchill," they greeted him as the savior of his country, the leader who had snatched Britain from the jaws of defeat, the man who more than any other had made this day possible. Three weeks after he had been hurled from power, the ovations comforted and reassured him. When he went to the palace afterward to congratulate George VI on his address in the House of Lords, it was evident that Churchill was "gleefully anticipating" the speech he planned to deliver in the Commons the next day. That evening, London, which had spent so many nights in darkness, became a city of light. Great buildings were floodlit, bonfires blazed, and fireworks streaked the sky. All across the city there was singing and dancing in the streets.

The celebrations were still going strong on August 16 when Churchill spoke in Parliament of the danger of a new war more terrible than any in the past. He gave thanks that the atomic bomb had brought peace to the world, but he cautioned that it would be up to men to keep the peace. He called the bomb "a new factor in human affairs" and emphasized that with the advent of such a weapon it was not just the survival of civilization that was imperiled, but of humanity itself.

Hours after Attlee proclaimed that the last of Britain's enemies had been laid low, Churchill pointed out that significant differences had already arisen with their Soviet ally about the state of affairs in eastern and central Europe. He noted the emergencc of police governments and he observed that it was "not impossible that tragedy on a prodigious scale is imposing itself behind the iron curtain which at present divides Europe in twain."

A fortnight after the conclusion of the Potsdam Conference, he lamented that instead of resolving the most serious questions, the three leaders had handed them off to a committee of foreign ministers, which was to meet in September and was gifted with less far-reaching powers. As he had at Potsdam, he perceived a unique opportunity in the fact that thus far only the US had the bomb. He warned that the time to get a settlement was during the three or four years that remained before any other power was likely to catch up. There was not an hour to be wasted, he cautioned, and not a day to be lost.

Despite the gravity of what Churchill had to say, he enjoyed parrying interruptions by some of the new Labour members. With a puckish grin he played up the irony of being the man to press for free elections elsewhere in the world when he had just been overwhelmingly defeated at home. Thumbs in lapels, he avowed his faith in democracy, whatever mistakes the people might be inclined to make. Speaking of the ideal of "government of the people, by the people, for the people" that he hoped to see implemented everywhere in Europe, he delighted listeners of all political persuasions with the self-mocking aside, "I practice what I preach!"

One young Labour member wrote in his diary afterward that Churchill's speech had been "a real masterpiece"; the *New York Times* called it "one of the greatest speeches of his parliamentary career." Still, on a day of merrymaking in the streets, his words of warning were fatally out of sync with the national mood. Part of the power of Churchill's iconic wartime speeches was the extent to which they captured the nation's mood and gave magnificent voice to the hopes and ideals of the people. By contrast, the August 16 address was a throwback to his speeches in the 1930s, which had taken the form of warnings that no one wanted to hear, particularly so soon after the carnage of the First World War. In 1945, eager to start a new life under Labour, the war-weary British did not want to be informed that their trials had only just begun.

Clementine Churchill exulted that her husband's "brilliant moving gallant" speech had been on a par with his best work, but she too was ready for a new life after the war. The Churchills had left Claridge's for their eldest daughter Diana's flat in Westminster Gardens, where they were to stay while Clementine worked on reopening their country house, Chartwell, in addition to readying a new house in London. The couple had settled on a brick house in a cul-de-sac at Hyde Park Gate, off Kensington Road, but the sort of life they intended to have there remained a point of contention.

While Clementine, aged sixty, saw Hyde Park Gate as a retirement haven, Winston emphasized its proximity to the House of Commons. He assured fellow Conservatives he would be able to get to the House in less than fifteen minutes. By the use of certain shortcuts, the driver could probably do it in seven.

In the course of Churchill's long and tempestuous career, the one person whose backing he had always been able to count on absolutely was his wife. Through thirty-six years of marriage, Clementine had never lost faith in what together they called his "star," and never wavered in the mystical conviction that Winston was destined to accomplish great things. Repeatedly, in the face of political disappointment, she had soothed his bitterness and encouraged him to carry on. Consistently, she had put him above their children's needs and her own. She once told him that if to help him or make him great or happy she had to sacrifice her life, she would not hesitate.

Clementine also saw his flaws and did not fear to point them out to him. She was nothing if not critical. Nonetheless, whatever he wanted for himself this imposing and formidable woman had learned in some sense to want as well. His trials had been her trials, and his enemies had been her enemies. At least, that had always been the case—until now.

The present situation was lonelier and more personally painful than anything the Churchills had experienced to date because, suddenly, husband and wife were in open and irreconcilable conflict about how to spend the rest of their lives. During the Second World War, Clementine had worried about Winston's health to the point that her own was affected. Still, as long as the Nazi menace remained she had found a way to accept that her husband must put himself at risk and that she might lose him in the process. After the war, she saw things differently. Winston was old and ill. She loved him and she wanted them to be able to enjoy the few years they still had together. As far as she was concerned, his wartime leadership had vindicated the decades when they had sometimes been almost alone in the world in believing in him.

Though Winston persisted in dwelling on unfinished business, Clementine was confident that he had fulfilled his destiny at long last and that he—no, they—had earned a quiet, happy retirement.

While he certainly did not always do as his wife suggested, he prized her judgment and political acumen. He was always eager to know what she thought, and he would grow annoyed if she refused to tell him. In early 1945, Clementine had counseled her husband to retire as soon as the war was won and to refrain from seeking reelection. She wanted him to leave office, but that did not mean she wished to see him defeated in Britain's first postwar general election. On the contrary, when he insisted he was not ready to be "put on a pedestal," Clementine supported his candidature unreservedly.

Still, her comment that the election loss might prove to be a blessing in disguise went to the heart of a new kind of sadness in their marriage. Much as Clementine ached for him in defeat, she earnestly believed that they would both be better off in private life. Much as it pained her to see him again feel rejected and unappreciated, there could be no denying that in some sense she had gotten what she wanted. From this point on, the burdens of the premiership would fall to others.

After Parliament went into recess on August 24, Churchill had nothing to absorb and distract him. It was almost worse that he had known the fleeting joy of preparing and delivering his big speech. The letdown was stunning in its ferocity. As he once wrote, he found it very painful to be impotent and inactive. In the emptiness of his days he brooded about the election, but even in private he could barely bring himself to criticize those who had cast him out. (Clementine, interestingly, was less forgiving of the British public.) In his view the people's right to choose their leaders was the very thing he had fought the war to protect, and he struggled to suppress his bitterness at what he could not help but perceive as their ingratitude.

He was mightily unhappy, and his wife observed that that made him very difficult at home. Clementine regretted that, rather than cling to each other in their sorrow, they seemed always to be having scenes. He fought with her, with her first cousin Maryott Whyte, with their son, Randolph (always an eager sparring partner), and with others. He complained about the food he was served; he protested at the lack of meat now that he had to endure the same physical shortages as other men; he imagined that the "gruff bearish" cousin, an impoverished gentlewoman who assisted Clementine in household matters, was intent on thwarting him at every turn. He wanted to have cows and chickens at Chartwell. "Cousin Moppet" maintained it would never work. Feathers flew. Clementine said she was sure it was her own fault, but suddenly she was finding life with Winston more than she could bear.

The people had spoken, and by any realistic assessment Churchill was never going to be prime minister again. Eventually, even he accepted that no one could go on being this miserable and that he had to come to peace with what had happened to him; but how?

III

SANS SOUCIS ET SANS REGRETS

Lake Como, September 1945

During the five-and-a-half-hour flight to Milan in a Dakota aircraft provided for his personal use by the Supreme Allied Commander for the Mediterranean, Field Marshal Alexander, Churchill pored over five years' worth of his wartime minutes. This was the torrent of dictated notes, consisting of comments, questions, and requests to individual ministers, to the chiefs of staff, and to others, by which he had brought his powerful personal impact to bear on every aspect of the conduct of the war.

Churchill had used his minutes the way an octopus uses its tentacles—to reach everywhere, to be in many places at once. Throughout his life, he had been constitutionally incapable of sitting back and letting others do what he usually believed he could better accomplish himself. Where problems existed, he was driven to grapple with them directly, so much so that at times even his admirers had been known to question his sense of proportion. He craved responsibility, which he once tellingly described to his mother as "an exhilarating drink." His minutes allowed him to be involved in anything that concerned the fight against Hitler, to shine a searchlight into the most obscure corners of the war effort, and not only to learn about, but also to manage details which other, less controlling personalities might have been inclined to leave to the judgment of subordinates. Now, all that power had fallen away from him.

Still, he had not brought printed copies of his minutes just to brood over what had been lost. Faced with the likelihood that his political career was at an end, Churchill insisted he could not simply be idle for the rest of his life. More and more, it seemed as if he was weighing the possibility of a memoir along the lines of *The World Crisis*, his highly personal multivolume chronicle of the First World War. Fellow Conservatives—most, apparently, with the ulterior motive of edging him aside as party leader—had suggested that he undertake to tell the story of the Second World War as only he could. At a time when Churchill was deeply upset that the election had called his record into question, a memoir held the distinct attraction of allowing him to defend his actions, both during the war and in the immediate aftermath, in the courtroom of history.

Could such an undertaking fill the vacuum in his life that had been created when he lost the premiership? Would a memoir be enough to absorb the energies of a man of Churchill's temperament? In part, that was what he was on his way to Italy to find out.

As Parliament was in recess until October 9, Alexander had offered him the exclusive use of the Villa La Rosa, the commandeered property above Lake Como which had served as the field marshal's headquarters in the war's final days. In anticipation of his stay there, Churchill had had bound copies of his minutes and telegrams specially prepared; these would form the spine of any autobiographical work. The Churchills' middle daughter, the redheaded Sarah, was with her father on the flight on September 1, along with his physician, his secretary, his valet, and a detective. He had wanted Clementine to come as well but she refused, explaining that she would be able to accomplish more in his absence. She too was exhausted and dejected, and felt that she would be unable to enjoy a holiday in the sun.

On the plane Churchill barely said a word to the others, but in the car afterward it became apparent that in the course of reading he had already seized on the narrative possibilities of one part of the Second World War saga. He spoke excitedly of the Dunkirk evacuation, testing the story, feeling for the drama. At the Villa La

Rosa, after he learned that one of the aides-de-camp assigned to him for the occasion had been at Dunkirk, animated talk of the episode continued over dinner. Seated at a huge green glass table in the ornately-mirrored and -marbled pale green oval dining room, Churchill interrogated the nervous twenty-four-year-old. How long had he waited on the beaches? What kind of vessel had rescued him? Churchill, in his enthusiasm, wanted to hear every detail.

Previously, he had been in such low spirits that Sarah had feared time would pass slowly and dully. Already, that was far from the case. But any hope that a change of scenery was all that it would take to cure her father was soon dashed. After dinner, Churchill put on a dark hat and coat over his white suit, and padded out onto the balcony in bedroom slippers to sit. As he puffed on a cigar, his interest in a memoir seemed to evaporate with the swirls of smoke. He insisted to his doctor that he was in no mood to write, especially not when the government was poised to take so much of his earnings. Suddenly, he was back to rehashing the election, brooding aloud about what had gone wrong and what might have been.

Early the next morning, the sun was warm and bright and a soft breeze rippled the lake as a tiny caravan assembled in front of the Villa La Rosa, which gave long views of villages and mountains on the opposite shore. An aide-de-camp loaded one of the cars with Churchill's painting apparatus. An elaborate lunch was packed in an accompanying station wagon. Through the years, Churchill had often sought relief, repose, and renewal through painting. He first picked up a paintbrush in 1915 after the loss of his position as First Lord of the Admiralty, at the time of the calamitous Dardanelles campaign, affected him so strongly that Clementine worried he would "die of grief."

Then, as now, he had been cut off in the midst of a great and urgent undertaking. Then, as now, it galled him to be deprived of control while the fate of the enterprise was still in suspense. Then, as now, he felt as if he "knew everything and could do nothing." Then, as now, at a moment when every fiber of his being was "inflamed to action," he was forced to remain "a spectator of the tragedy, placed cruelly in a front seat."

In a period when dark broodings about his predicament had allowed him no rest, painting had come to his rescue. Thirty years later, his daughter and others in the group, not to mention Churchill himself, were hoping it might do so again. The vehicles were packed and ready to go by 10 a.m., but Churchill did not enter the open yellow car until almost noon.

They drove along the lakefront while Churchill scouted for what he liked to call a "paintaceous" scene. Before long, he announced that he was hungry, so the procession halted and a table was set up. About twenty Italian peasants formed a circle around the English travelers and watched them eat and drink. At length, the Churchill party drove over the mountains to Lake Lugano. It was late afternoon before he found a view that pleased him. His easel, canvas, paints, and brushes were laid out, along with the tiny table he liked to have nearby for whisky and cigars. The paints were arrayed on a tray fitted to stand slightly above his knees. The brushes went in a ten-inch-high container. Finally, wearing a white smock and a straw sombrero, Churchill settled into his cane painting chair and began to work.

When his sister-in-law Gwendeline Churchill, known as Goonie, introduced the middle-aged Churchill to painting in 1915, he found that he needed only to concentrate on the challenge of transferring a scene to his canvas in order to put politics and world problems out of his thoughts. For a man who worked and worried as much as he did, the discovery was a revelation. His private secretary later said that it was as if a new planet had swum into his ken.

Then and on many subsequent occasions, though not during the Second World War when the magnitude of his burdens allowed no interruption, the balm of painting healed Churchill both mentally and physically. He painted in rapt silence. As he focused on a composition, all of his cares and frustrations appeared to vanish. He reveled in the physical and tactile aspects of the process, from the "voluptuous kick" of squeezing the fragrant colors out of their tubes, to the capacity for building the pigment "layer after layer," to the wondrous ability to scrape away one's mistakes with a palette knife at the end of the day. When he inspected a finished painting, he was known not just to look but also to touch the surface of the canvas, caressing the whorls of dry paint with his fingertips.

Churchill theorized that when he painted, the use of those parts of the mind which direct the eye and hand allowed the exhausted part of his brain to rest and revive. A change of scenery alone would not have sufficed, for he would still be condemned to think the same thoughts as before. Nor would activities like reading or writing, for they were too similar to the sort of work that had worn him out in the first place. Nor would it help simply to lie down and do nothing, for the mind would keep churning. Painting offered a complete change of interest. It was not that his thoughts stopped; he was thinking, to be sure, but about matters other than those that had been preoccupying him.

At Lake Lugano in 1945, painting again seemed to work its magic. Eyeglasses partway down his nose, Churchill paused at intervals to push back his straw hat and wipe his forehead, but otherwise he labored continuously, utterly engrossed. Another group of Italians, mostly children no more than twelve or thirteen years old, sat on the ground and observed. By the time he put down his brush at last and the spell was broken, five hours had passed and it was early evening. Later, Sarah was pleased to hear her father exclaim, "I've had a happy day." As she reported to Clementine, she had not heard him say that "for I don't know how long!"

Churchill continued to paint in the days that followed. The only drawback was that at his age too much sitting threatened to stir up the hot lava of his indigestion. In the evenings, he would prop up his canvases in the dining room and appraise them during dinner. He transformed his huge bathroom, which had mirrors on every wall, into a studio with makeshift easels, and he would stare at works in progress while he soaked in a marble tub. He rejoiced that in Italy he felt, as he had not in many years, as if he were entirely out of the world. At home he was an obsessive reader of newspapers, which he marked up with slashes of red ink before dropping them on the floor for someone else to collect. Here he saw no newspapers for days at a time. When they were delivered, he claimed to be so busy with his painting that he hardly had time to read them.

In this spirit Churchill was soon insisting he was glad to have been relieved of responsibility for how things turned out after the war. He claimed as much in separate conversations with Moran and with another physician, who came to the villa to fit him with a truss. He wrote to Clementine of his own steadily growing sense of relief that others would have to deal with the problems of post-war Europe. And he told Sarah one evening, "Every day I stay here without news, without worry I realize more and more that it may very well be what your mother said, a blessing in disguise. The war is over, it is won and they have lifted the hideous aftermath from my shoulders. I am what I never thought I would be until I reached my grave 'sans soucis et sans regrets.'" In a good deal of this, Churchill was probably trying to convince himself as much as anyone else of his change of heart. Certainly he had gone through this very process at the time of the Dardanelles disaster, pretending to be content with the loss of high office when in fact he was waiting and hoping for an opportunity to regain power and influence.

Unlike his air of calm acceptance, the healing effects of his artist's holiday were no pose. Churchill had long been blessed with remarkable powers of recuperation. At Lake Como, his absorption in something other than personal and professional issues allowed those powers to kick in. At the end of eighteen days he seemed so much better physically and mentally that he decided to extend his trip, sending his doctor back to England along with Sarah and nine finished canvases. Accompanied by his remaining entourage, Churchill drove along the Italian and French Riviera in search of new scenes to paint.

On his first day out he motored for four hours through ravishing countryside to Genoa. He arrived after nightfall to find the British officer who was in charge of the area ensconced at the Villa Pirelli, an "incongruous" mix of marble palace and Swiss chalet perched on a rocky bluff above the sea. Churchill's host marveled at how healthy and vigorous he looked after so many years of war. But admiration turned to alarm when Churchill proved rather too active for his host's comfort. In the morning, Churchill insisted he wished to swim despite the fact that the clear, pale green water below was said to be somewhat rough and the bathing place rocky. Refusing to be talked out of his plan, he climbed down nearly a hundred steps, followed by his valet carrying a massive towel. Soon, Churchill had doffed his silk dressing gown and bedroom slippers and was splashing about, porpoise-like, enjoying himself immensely. At the end of the session, an awkward logistical problem required the poor beleaguered host to push Churchill's boyishly pink-and-white, five-foot-six, 210-pound figure up from the water while an aide-de-camp tugged from the shore.

After two days in Genoa, Churchill and company proceeded to the half-empty Hôtel de Paris in Monte Carlo, where he dined lavishly on a veranda overlooking the casino and confronted a stack of newspapers from London. He had seen some press at Lake Como, but in that setting the information had struck him as oddly remote. Now, revived in body, mind, and spirit, he took in the first British reports of discord and deadlock at the Council of Foreign Ministers, which had been meeting in London in his absence. Molotov had thrown every obstacle he could think of in the path to progress (even so, as was later discovered, Stalin had berated him in secret messages for being too soft and conciliatory). For many observers in Britain and elsewhere, the talks' failure amounted to a first disconcerting glimpse of the sharp divisions that had already emerged between the Western democracies and their wartime ally the Soviet Union. The fiasco came as no surprise to Churchill. He had predicted as much in the House of Commons on August 16, when he publicly lamented the handing off of the most serious questions at Potsdam, questions the heads of state themselves ought to have settled. Once again, his warnings about events in Europe were starting to come true.

When he moved on to Cap d'Antibes, where he stayed at a fully staffed villa on loan from General Eisenhower, he wrote to Clementine in a voice markedly different at times from that of his letters from Italy. Previously, he had claimed to be interested solely in painting and to have little appetite for news of the outside world. Now, he spoke of how certain he had been that the foreign ministers' talks would fail, of his understanding that the Soviets had no need of an agreement as they actually welcomed the chance to consolidate themselves in nations already in their grasp, of his concern that so little was known about what was happening to the Poles, the Czechs, and others trapped behind the iron curtain, of his sense that the future in Europe was full of "darkness and menace," and of his feeling that there would be no lack of subjects to discuss when Parliament reconvened. Clearly, this was the letter of a man ready to reengage.

Churchill had gone to Italy in the hope of coming to peace with the people's decision. He had tried very hard to concur with his wife that the loss of the premiership was indeed a blessing in disguise. At last, he found he could do neither. The threat of another war was too great. His confidence that he was the man to prevent it was too strong. For better or worse, it simply was not in his character to remain detached for long.

When Churchill returned to Britain on October 5, 1945, his family understood that he had made an important decision while he was abroad. Once again, in defeat he would be defiant. Whatever the obstacles, he intended to fight on. He refused to retire.

IV

OLD MAN IN A HURRY

London, October 1945

It was one thing to decide to fight on, quite another to stage a political comeback in his seventies.

During his first week in London, Churchill was a whirling dervish of activity, leaving no doubt in anyone's mind that he meant, and retained the capacity, to lead. He set policy with his Shadow Cabinet. He cut a lively figure on the opposition front bench. He offered the first opposition motion and he directed all Conservatives to be present the following week when he assailed a bill to prolong government controls on labor, rations, prices, and transport for five additional years. Parliamentary commentators noted his bronzed, robust appearance, and King George remarked privately that Churchill had returned from Italy and France "a new man." As if he had energy to spare, Churchill capped off a busy week by attending a Friday evening performance of Oscar Wilde's *Lady Windermere's Fan.*

By the next morning, however, his efforts began to unravel. While her husband was at Lake Como, Clementine Churchill had worried that in his passion to transfer a scene to the canvas he might labor on, oblivious to the chill of the evening air. Given his medical record, there was always anxiety that were he to catch cold it could escalate into pneumonia. As feared, he returned from the South of France with a cold. Despite promises to be careful, he largely ignored it. By Saturday, he had lost his voice. By Sunday, a statement went out that he was confined to his house on doctor's orders due to an inflamed throat.

Churchill had rallied the troops, but in the end he would not be there to lead them. Instead, to his frustration, he spent the week in his sickbed unable to speak. At a moment when he had been eager to fashion an image of vitality, press reports brought up his prolonged bouts of pneumonia during the Second World War, his impending seventy-first birthday, and the undeniable fact that for a man of his years a tiny cold could prove to be a very big deal. In view of his comeback plans it was all a bit of a disaster, but as a friend once said, Churchill "produced his greatest efforts in disaster." Adversity tended to stimulate him.

While he was abroad, a stack of invitations to speak had accumulated at Hyde Park Gate. One request in particular fired Churchill's imagination by appealing to his sense of drama. F. L. McLuer, the president of Westminster College in Fulton, Missouri, asked him to lecture on international affairs. The proposal dovetailed with Clementine Churchill's wish that they spend part of the winter in Florida for his health. Still, McLuer's letter is likely to have been of little interest had it not been for an addendum scribbled across the page: "This is a wonderful school in my home state. Hope you can do it. I'll introduce you. Best regards, Harry Truman." Even Truman's seconding of the invitation would have made little impact on Churchill without the offer, however casual and offhand, to introduce him.

Churchill instinctively grabbed on to those three words as if they were a lifeline, and he refused to let go until he had used them to hoist himself back up onto the world stage. Truman's presence on the same platform would call world attention to his message about the looming Soviet threat in a way he could never hope to achieve himself. In his present circumstances, it would mean everything to Churchill, a defeated politician after all, to be able to borrow and bask in the American leader's power.

As Churchill crafted his reply he went significantly beyond an attempt to formalize Truman's commitment. He tried to draw the president further into the picture, to suggest that Truman had intended a good deal more by his words than he probably had. (In fact, Truman as yet had no real investment in the visit. An intermediary, a Westminster alumnus, had solicited his involvement in the invitation. Truman had merely added his hasty endorsement and passed the letter on as a favor to a friend.) Though it was McLuer who had written to him, Churchill cut the college administrator out of the loop by addressing his letter, dated November 8, 1945, directly to Truman. He wrote as if he had in hand an official presidential invitation to speak under Truman's "aegis." Careful to refer to his understanding that Truman planned to introduce him, Churchill insisted it would be his "duty" to come to the US and do as the president requested. He pledged to Truman that the Fulton speech would be his only public address in America "out of respect for you and your wishes."

Churchill was unquestionably distorting the tenor of Truman's message, and his decision to point out that he had praised Truman in the House of Commons the previous day was also risky. At Potsdam, Truman had complained in his diary of what he perceived as Churchill's efforts to soft-soap him. Nevertheless, having cleared the speaking engagement with his successor, Churchill sent off his answer via Attlee's secretary, to be hand-delivered when the American and British leaders met presently to discuss atomic policy and related matters in Washington.

While Churchill waited for Truman to respond, he went to Paris and Brussels for a week to speak and be feted. His painting holiday had helped him regain perspective and confidence after the election and he was happy once again to receive honors. Now, with an eye toward a comeback, it suited him to shift public attention back to his war triumphs.

In Brussels, adoring crowds waited for hours to catch a glimpse of him. They fought their way past police and tossed flowers at his car. A girl managed to hurl herself onto the running board and kiss him, and an old woman was heard to declare that now that she had seen Churchill she was ready to die. He was made an honorary citizen and proclaimed "the savior of civilization." Hailed for his war leadership, Churchill missed no opportunity to showcase his achievements in the run-up to war as well. When he told a joint session of the Belgian Senate and Chamber on November 16 that had the Allies moved to stop Hitler early on, the Second World War ("the unnecessary war") would probably never have had to be fought in the first place, he was reminding people that he had been right in the 1930s and letting them know he was right now.

In contrast to his rapturous reception abroad, there were no cheers for Churchill when he returned to London on November 20. Immediately, he faced a new challenge to his leadership. This time the malcontents were younger parliamentarians who mocked their tired elders in the Conservative Party as "Rip Van Winkles," content to sleep through the socialization of Britain. To the young Tories' outrage, Churchill had been absent from Parliament on November 19, resting at Chartwell after his trip, when Labour unveiled further nationalization plans. The party's number two man, Anthony Eden, had been missing as well.

At a meeting of the backbenchers, Churchill slouched in a red leather armchair for an hour and a half, but he might as well have been enduring a slow stretch on the rack as his juniors by many years criticized his leadership. At the time of the general election, Churchill's belligerence had landed him in trouble; now the complaint was that he was not belligerent enough. The man of blood had gone anemic. The young people wanted him to set off a debate in the House of Commons on the broad matter of nationalization by introducing a motion of government censure.

No one can have wanted to turn out the government more than Churchill. No one can have had greater reason to be impatient. He was truly, as his father had said of the prime minister William Gladstone, an "old man in a hurry." Still, he protested, the timing was all wrong. The Attlee government had only been in power for a few months and it was too soon to argue that they had failed. Like it or not—and Churchill did not like it—the opposition had little choice but to wait upon events. The Conservatives needed to let some time pass and give things a chance to go wrong. Far from benefiting Conservatives, Churchill argued, a premature confrontation would spotlight Tory weakness, allowing Labour to emerge even stronger than before.

Churchill suggested that when his critics had had more experience they would see that he had been right, but they were unyielding. At last, he reluctantly consented to go on the warpath against Attlee; it was either that or allow the charge to stand that somehow he had lost the will to fight.

On the evening of November 27, three days before his seventy-first birthday, Churchill placed a motion of censure before the House, which claimed that the government had focused on long-range nationalization plans at the expense of the people's immediate postwar needs. Churchill filed the motion without comment in expectation of a full-dress debate the following week.

Robert "Rab" Butler, Churchill's wartime minister of education, established the tone at the next day's Conservative Central Council meeting in London. The pale, balding, pouchy-eyed Butler introduced Churchill as the "Master Fighter." Churchill's mockery of socialist ministers elicited peals of delight, and when he slowly, mischievously flapped his arms to help listeners visualize "the gloomy vultures of nationalization" hovering over Attlee's Britain, the hall echoed with appreciative laughter. Delegates from throughout Britain insisted they had never known Churchill to be in better form.

He met a less enthusiastic reception in the House of Commons. Labour shot down Churchill's motion—by this time, despite its genesis, it was very much identified in the public mind as his motion—by a vote of 381 to 197. But then, he had expected it to fail. What he could not have expected was the wit and ferocity of Attlee's counterattack. Churchill was known to view his successor as "a sheep in sheep's clothing," but there was nothing sheepish about Attlee's devastating performance on December 6.

Clementine Churchill watched from the gallery, and more than a hundred politicians had to stand or squat on the floor for want of seats, as the small, spare, fidgety Attlee, who had a reputation as a lackluster speaker, gave what was widely received as the best speech of his parliamentary career to date. Attlee made Churchill seem ridiculous for asking why a government that had been elected to carry out a socialist program did not carry out a Conservative program. He avowed that Britain disliked "one-man shows" and he characterized the motion of censure as nothing more than "a party move by a politician in difficulties."

Every time Attlee scored a hit—and there were many—the Labour benches roared. His plush pink target looked on in silence. Churchill made a point of rising above the abuse. Still, that Attlee had out-debated him was a blow to his prestige. Soon, it looked as if it might even have been a knockout, and the talk in political London was that Churchill might be preparing to step aside.

In fact, that was the last thing he meant to do. While Churchill had been managing the unrest in his party, Truman had officially confirmed his offer to introduce the Fulton speech. Since then, Churchill had been back and forth with Washington to press for a firm date, to ask that the event be announced simultaneously from the White House and in London, to urge Truman to make public his endorsement of the invitation, and to express a wish for talks between the president and himself. Ironically, when Truman granted all of these requests, the news of Churchill's impending trip, to speak in Missouri and to enjoy a rest in Florida with Mrs. Churchill, sparked new rumors of resignation.

Speculation was rife that Churchill's willingness to leave Britain at a time of deep division in the Conservative Party meant that he intended to give up the leadership upon his return. There were reports in the world press that he was traveling to Florida on doctor's orders and that the state of his health might soon force him to retire. Meanwhile, mindful of the havoc that had ensued when both he and Eden were missing from Parliament on November 19, Churchill reassured a large gathering of opposition members that Eden was set to lead in his absence. Instead of allaying fears, however, his comments provoked upset in certain Conservative quarters.

Eden enjoyed broad support in the party, but if indeed Churchill was preparing to hand over, not everyone was pleased with the prospect of power passing to Eden. His critics dismissed him as a lightweight who possessed more style than substance and who had risen only because so many of the best young men of his generation had perished in the First World War. In a public challenge to received wisdom about the succession, the *Evening Standard*, which was owned by Lord Beaverbrook, questioned whether Eden was quite up to the task. There followed a round of press comment, both at home and abroad, about Rab Butler and other possible successors should Churchill retire.

As 1946 began, representatives of fifty-one countries gathered in London for the first United Nations General Assembly. On January 9, final preparations were under way at St. James's Palace for that night's state banquet on the eve of the historic session when the Churchills sailed for America. Their giant liner, the *Queen Elizabeth*, which had delivered Eleanor Roosevelt and other members of the US delegation four days previously, was part of the effort to repatriate nearly two million American and Canadian troops that had begun after the surrender of Germany. On the present westward crossing, more than twelve thousand Canadians were finally on their way home. The day before they reached New York, Churchill addressed the troops over the ship's loudspeaker system. In the course of speaking to them of their future, the old warrior offered some hints about how he saw his own.

As the young men prepared to begin new lives after the war, Churchill promised them that the future was in their hands and that their lives would be what they chose to make them. The trick, he told them, was to have a purpose and to stick to it. He recalled that the previous day he had been standing on the bridge "watching the mountainous waves, and this ship—which is no pup—cutting through them and mocking their anger." He asked himself why it was that the ship beat the waves, when the waves were so many and the ship was one. The reason, he went on, was that the ship has a purpose while the waves have none. "They just flop around, innumerable, tireless, but ineffective. The ship with the purpose takes us where we want to go. Let us therefore have a purpose, both in our national and imperial policy, and in our private lives."

Some people at the time interpreted those remarks as Churchill's "farewell to politics." In retrospect, they appear to have been anything but that. Far from being inclined to shut down his political life, Churchill, though he too was no pup, was about to restart it.

V

THE WET HEN

St. James's Palace, 1946

A cold rain pelted London on the night of Britain's first state banquet since 1939. Inside St. James's Palace, crackling wood fires perfumed the air. Servants wore prewar red-and-gold and blue-and-gold liveries, and royal treasures that had been stored away for the duration of the war were once again on display. Candles twinkling in gold candelabra illuminated a banquet table set for eighty-six with heavy gold plate. As each of the fifty-one chief UN delegates and other guests entered, they were taken to a cavernous, tapestry-lined room where they were presented to the King. The fifty-year-old George VI wore the uniform of an Admiral of the Fleet. The Colombian delegate responsible for overseeing the preparations for the first General Assembly sat at his right, and the Belgian who was expected to be elected its president the following day sat to the King's left. Among the topics dominating the delegates' conversation was who would be appointed to the post of secretary general.

Hours after Churchill sailed, Anthony Eden arrived at the UN dinner for what would be his first public appearance as deputy leader of the opposition. For at least a week he had been fuming at the prospect of being left in charge, as he complained to Cranborne, "rather like a governess on approval." Unlike Churchill, he lacked the stomach to prove himself again. Eden believed that as foreign secretary, as well as leader of the House of Commons, he had demonstrated his abilities and should not have to endure another round of tests. Had not a decade passed since the prime minister Stanley Baldwin made him Britain's youngest foreign secretary since the mid-nineteenth century? Had not Churchill singled out his experience and capacity when he anointed him heir apparent?

There had been a time before the war when Eden struck many of the anti-appeasers as a more viable candidate for prime minister than the pugnacious, provocative, unabashedly and carnivorously ambitious Churchill. There had been a time when Cranborne, Eden, and others in their circle had barred Churchill from their meetings because they thought him unstable, untrustworthy, and unsound, and because they feared he would dominate their discussions and corrupt their cause by involving them with the adventurers who formed his claque. There had been a time when Eden's determination to bring Conservatives together and to formulate a unified Tory position on the Fascist threat had seemed much more sensible and appealing than Churchill's willingness, even eagerness, to split the party asunder.

Cranborne believed that when Churchill became prime minister he never really forgave the Edenites for shutting him out. Close observers would long suspect that however highly and affectionately Churchill spoke of Eden, he truly "despised" his second-in-command. One could never be sure: when Churchill ostentatiously referred to Eden as "my Anthony," was that a note of contempt in his voice? Nevertheless, from early on the matter of the succession in general and of Eden's claims in particular had been prominently in play. At the outset of his premiership, Churchill had spoken of his intention to resign at the end of the war to make room for younger men. In 1940 he told Eden that he regarded himself as an old man and was not about to repeat the prime minister Lloyd George's error of attempting to carry on after the war. On various occasions and in various ways he made it clear that he wanted Eden to succeed him.

As the war dragged on, it seemed as if Eden would not have

to wait for the peace after all. When there was broad dissatisfaction with aspects of Churchill's leadership and the progress of the war, when the old man was gravely ill, and when there were fears he might soon die, Eden had had reason to believe the handover would occur at any moment. Both verbally and in his letter to the King, Churchill spelled out his wish that should anything happen to him Eden would take his place.

Despite Churchill's assurances to Eden that it would not be long before the younger man took control, somehow that golden day always failed to arrive. There were persistent grumblings in certain quarters that the prime minister was "losing his grip" (Sir Alexander Cadogan 1942) and "failing fast" (Field Marshal Sir Alan Brooke 1944), that he had grown too old, sick and incompetent, and that he really ought to "disappear out of public life" (Brooke) before he damaged both his reputation and the country; but still Churchill managed to endure. As a friend of Eden's later said, "Waiting to step into a dead man's shoes is always a tiring business, but when the 'dead' man persists in remaining alive it is worse than ever." Ironically, in 1944 it was Churchill who took on Eden's duties in addition to his own, when the ill, exhausted heir apparent, twenty-two years his junior, needed to go off for a rest.

When Churchill battled to retain the premiership in Britain's first postwar election, Eden was perhaps only being human when he discovered that he could not stifle "an unworthy hope that we may lose." And when he did get what he had guiltily wished for, he confided to his diary that while history would dub the British people ungrateful for having dismissed Churchill, perhaps they had, in reality, only been wise. Believing that the Tories would be out for ten years, Eden spoke to friends of his fervent desire to lead the opposition and mold the party for the future, but he also voiced concern that Churchill would insist on holding on to the job—"and get everything wrong." By the time six months had passed, Eden's fears, at least those about Churchill's intentions, seemed to have been realized.

As far as Eden could tell, at the start of 1946, Churchill had not even contemplated the possibility of retirement. Eden whined to Cranborne that Churchill meant to go on "forever." He was sure that the Conservative chief whip James Stuart and the party chairman Ralph Assheton were encouraging Churchill to hold on to the leadership for as long as possible in the interest of putting off the "evil day" when Eden took over. And even if Churchill were miraculously to step down, Eden was no longer confident that the party leadership, not to mention the premiership, would ever be his. He worried about being displaced by the likes of the forty-two-year-old Rab Butler or the fifty-one-year-old Harold Macmillan, though neither man was generally regarded as ready to lead. He also worried about the impatient young Tories who, when they mocked the party's Rip Van Winkles, meant the second-in-command and other venerable Conservatives no less than they did Churchill. To make matters worse, Eden's tumultuous personal life threatened to bar him from the premiership for good. His marriage was in tatters; Beatrice Eden wanted to marry her American lover. A divorce could sink Eden's political dreams. Had he worked and waited all this time—for nothing?

Eden's complaint was not that he could have been a contender. It was that he had been one for too long. At a moment when he felt "fed up with everything," the prospect of a new job suggested a way out of the succession trap. Even as Churchill had been using Eden to allay concerns about his own impending absence in America, Eden had been hoping he might soon be in a position to bolt. Nothing had been settled and other names were still prominently in play, but on January 2, 1946, Ernest Bevin confirmed to Eden that he was a candidate to become the United Nations' first secretary general. Following their talk, Eden let it be known at the palace that he was "anxious" to be considered.

Characteristically, he was not without ambivalence. Eden had a lifelong tendency to vacillate that had prompted Lady Redesdale, the mother of the Mitford sisters, to dub him "the wet hen." In the present instance, he seemed to be scurrying in all directions at once. Eager as he was to escape to the UN, he hesitated to abandon his prime ministerial ambitions after all that he had done and endured to realize them. By turns he insisted that he longed to extricate himself from the rough-and-tumble of British politics and he vowed to return to lead his party when Churchill was gone at last. Eden's former boss, Stanley Baldwin, warned that if he joined the UN, Butler was likely to claim the Tory leadership; once Eden made the move, there would almost certainly be no coming back.

Cranborne huffed to Conservative colleagues that if Eden took the UN job he was "through with him." Typically, however, Cranborne assumed a very different posture in conversation and correspondence with Eden himself. Rather than threaten Eden, he flattered him. In his most narcotic tones, Cranborne encouraged Eden in the belief that he was indispensable to the party's prospects. He maintained that only the designated heir could keep Conservatives together and that only he could lead them to victory in a new general election. For Cranborne, the UN episode was a flashback to the offstage tempest three years previously when Eden, already maddened by Churchill's staying power, had considered becoming viceroy of India. At the time, Eden had assumed that his position as Conservative heir apparent would be waiting whenever he saw fit to return. Now, as then, Cranborne, acting in his accustomed role of providing "the backbone to Eden's willow," worked hard to disillusion him. In the process, Cranborne may merely have substituted one illusion for another. He reassured Eden that Churchill's day was finally over, that Churchill now belonged to the past, and that even he was bound to find this out. Cranborne made the case Churchill had often made himself: that Eden needed only to be patient and wait a little longer as number two.

Eden was a figure of stark contradictions. As a diplomat he was a nimble negotiator gifted with an ability to mitigate tensions and always to seem cool and composed. As a man he was also vain, touchy, and hysterical. Alcohol brought out the worst in him. When Eden arrived at St. James's Palace on January 9, the King's private secretary, Sir Alan Lascelles, guessed that he had been drinking. Immediately, Eden ripped into two of the King's equerries as they showed him to his place at the banquet table in the William IV Room. Loudly complaining that he had been seated next to the head delegate from Nicaragua, the would-be leader of the new world peacekeeping organization made no secret of his conviction that he rated a more important dinner partner. He also seemed unhappy that he had been placed in the vicinity of Attlee.

Dinner began; the Krug 1928 champagne and other wines from the cellars at Buckingham Palace flowed; and George VI, a slight man with prominent teeth who suffered from a nervous stammer that tended to affect him when he addressed the public, spoke of the momentous tasks facing the delegates and of the need to put petty, selfish concerns aside in the interest of making the UN a success.

After dinner, the company moved to the Queen Anne Room, where the King planned to talk individually with certain guests. He was especially keen to speak to the dark-eyed, sallow-faced Russian, Andrei Gromyko (said to call to mind "a badger forced into the daylight"), to urge that the wartime contact between London and Moscow not be lost. George VI's press secretary, Lewis Ritchie, brought over each of the chosen delegates, and photographs were taken at the King's request. The bright flashes provoked a new hissy fit from Eden. Using filthy, abusive language, he protested to Ritchie that the camera lights were bothering him.

Butler, one of a small number of opposition members present, sprang forward to apologize on Eden's behalf, pointing out, in case anyone had failed to notice, that the man he hoped to replace as heir apparent had had too much to drink.

The next day, Lascelles drafted a stinging letter of rebuke to Eden. Realizing that he had "made an ass" of himself, Eden, before he heard from the palace, wrote an abject letter of apology. Not only had he sabotaged his candidacy for the UN job, but he had also provided ammunition to those who questioned his capacity to lead the Conservative Party.

As it happened, Churchill was asked for his thoughts on both the secretary generalship and the Tory leadership when the *Queen*

Elizabeth docked in New York on the evening of January 14. Flags whipped in the frosty Hudson River winds and a US Army band struck up "Hail, Hail, The Gang's All Here" as he descended the gangplank. Observed by the Canadian troops, whose heads stuck out of many of the ship's portholes, Churchill made one of his dainty half-bows to a large crowd of press and American, British, and Canadian officials.

Afterward, in a heated waiting room on the pier's upper level, he thanked reporters for coming out on such a cold night and gamely took their questions. Clementine Churchill, swathed in black furs, helped with any words he failed to hear. In the course of bantering with reporters, Churchill addressed topics that had been the subject of speculation and gossip in London for weeks. His remarks were of particular concern to certain personally interested parties at home.

Did Churchill plan to retire from active politics? "I know of no truth in such reports," he fired back. Was he going to hand over the Tory leadership? "I have no intention whatever of ceasing to lead the Conservative Party until I am satisfied that they can see their way clear ahead and make a better arrangement, which I earnestly trust they may be able to do." Was Churchill prepared to serve as the first UN secretary general? This question appeared to puzzle him. Churchill knew that Eden wanted the job, and before he left the country he had "strongly" advised him to accept were it to be offered. (Eden, for his part, assumed Churchill wished to see him settled elsewhere so he would feel easier in his mind about staying on. Churchill similarly had counseled Eden to take the viceroyship in 1943 on the explanation that he hoped "to go on some years yet.") At the time of the press conference in New York, Churchill had no idea that Eden had already torpedoed whatever chance he might have had to go to the UN. It was only now he discovered that one day previously in London some of the South American delegates had put forth the name "Winston Churchill" as the latest candidate for the post. After a second's reflection, he swatted the question aside. "I never addressed my mind to such a subject."

The following day in London, Eden chaired a meeting of the Shadow Cabinet. Cranborne hovered about Eden to be sure he made better use of Churchill's absence. In the wake of Eden's suicidal performance at the state banquet, Cranborne was pleased to see him act calmly and effectively to consolidate his position in the party. Where Eden had bristled at suggestions that he had yet to prove himself fully, Cranborne saw the deputy leadership as a huge opportunity for their side. Whatever assurances Cranborne had previously offered to Eden in the interest of dissuading him from accepting the post of UN secretary general, Cranborne did not really believe that Churchill would readily hand over any time soon. He did, however, hope that if the deputy leader performed well, Eden would be in a strong enough position to push Churchill out when the old man came home.

VI

WINNIE, WINNIE, GO AWAY

Miami Beach, Florida, 1946

Seated beside a bed of red poinsettias near the pink brick seaside house his wife had arranged to borrow from a friend, Churchill contentedly scanned the coconut palms overhead in search of a "paintaceous" angle. His tropical-weight tan suit fit snugly across his stomach. The deep creases radiating from the center button, which looked as if it was about to burst, testified that he had grown thicker since he acquired the suit in North Africa during the war. In the white patio chair beside him, Clementine Churchill wore one of her customary headscarves, big round white-rimmed sunglasses, and wrist-length white gloves. After the bone-chilling cold they had had to endure in New York harbor and the rain-splashed train windows en route through Virginia and the Carolinas, she proclaimed the intense heat and sunshine on the day they arrived in Miami Beach "delicious." Churchill had lately suffered his share of colds and sore throats, and in keeping with his wife's wishes he intended to rest and to enjoy the good weather in Florida. Still, from the outset the couple had contrasting perspectives on their stay. She saw their holiday as an end in itself, he as a chance to get in shape for the main event in Missouri.

The next morning, the Churchills were unhappily surprised. The sky had darkened and the temperature had plummeted. There followed a day and a half of shivering cold and rustling palm fronds until the afternoon emergence of the sun prompted Churchill to rush off with his painting paraphernalia. He worked for hours in the shade on a picture of palms reflected in water. Despite the knitted afghan which Clementine draped around his shoulders when she brought him his tea, he caught another cold and was soon running a slight temperature. The episode was exactly the sort of thing they had come to Florida to avoid. At a time when he was supposed to be gearing up for Fulton, the usual concerns about pneumonia plunged him into a fit of agitation. For all of his philosophy, he always found it maddening when illness threatened to get in the way of his great plans. Friends affectionately called Churchill the world's worst patient. This time, he alternated between insisting he wanted no medicine and taking several conflicting remedies all at once.

His fever broke after thirty-six hours. The perfect weather resumed and Churchill was able to paint again and to swim in the ocean. Welcome news arrived in the form of a message from Truman that he would soon be on holiday in Florida and would be happy to dine with Churchill on the presidential yacht. The prospect freed Churchill from the need to brave any more bad weather were he to have to fly north to confer with Truman. In the meantime, Truman sent a converted army bomber to transport the Churchills to Cuba for a week of painting and basking in the sun. The president and the former prime minister were set to meet after that, but when Churchill returned from Havana he discovered that Truman had had to cancel his vacation because of the steel strike. Churchill insisted he would fly to him the next day.

The exceptionally rough five-hour trip proved to be an ordeal. Churchill was finishing lunch when the B-17 bomber passed into a sleet storm above Virginia. Suddenly, plates and glasses pitched in all directions and Churchill was thrown against the ceiling. Not long afterward, the aircraft landed safely amid a swirl of ice pellets. Churchill rose amid the shattered glass that covered the cabin and relit his cigar by way of composing himself. He descended the steps at National Airport beaming and waving his hat to Lord Halifax and other official greeters as if he had just enjoyed the most tranquil of flights. After he had bathed and dined at the British embassy, he was off to the White House to meet Truman for the first time since Potsdam.

When Churchill last saw him, Truman had recently inherited Roosevelt's unrealistic perception of Stalin, as well as his predecessor's tactic of dissociating himself from Churchill in an effort to win the Soviet leader's confidence. Accordingly, Truman had had little use for Churchill's perspective or advice. By early 1946, however, Moscow had given the president reason to reconsider. A series of speeches in January and February by Molotov and other of Stalin's lieutenants warning of the peril of an attack from the West had culminated, the previous day, in a bellicose address by Stalin himself. A translation appeared in American newspapers on February 9, the day Churchill flew into Washington. Stalin's enunciation of a tough new anti-West policy was a throwback to prewar Soviet attitudes. Immediately, as Halifax pointed out, the speech had the effect of "an electric shock" on the nerves of a good many people in Washington. Could this possibly be the wartime ally with whom they had been looking forward to close future cooperation?

In part, Stalin's confrontational tone had its origins in a two-month holiday he had taken starting in early October 1945. While the ailing, exhausted Stalin rested near Sochi at the Black Sea, he had left Molotov in charge of daily affairs at the Kremlin. The arrangement set off a chain reaction of rumor and gossip in the international press. By turns, Stalin was reported to be contemplating retirement, about to hand over to Molotov, and nearly or already dead. There were news profiles of Molotov and some of the other possible contenders should a fully-fledged succession struggle erupt on Stalin's demise. Though he was supposed to be resting, Stalin obsessively pored over a dossier collected under the title "Rumors in Foreign Press on the State of Health of Comrade Stalin." References to the second-in-command's ever-expanding prestige both at home and abroad fired Stalin's suspicions. Was Molotov behind the reports? Why had he not censored such material? Was the anointed heir using Stalin's absence to consolidate his position?

The rumors about Stalin's health had also distressed Churchill, who continued to hope that he might one day face him across the conference table and pick up where they had left off at Potsdam. Churchill therefore had been greatly relieved when the US ambassador in Moscow, Averell Harriman, announced that he had visited Stalin's seaside retreat and found the Soviet leader in good health. In the House of Commons on November 7, 1945, Churchill had expressed gratitude that Stalin was well, offered some kind words about his leadership, and voiced a wish that the bond that had developed between their two peoples during the war be allowed to continue in peacetime. On the face of it, Churchill's remarks were innocuous. Nonetheless, when Molotov directed that they be published in *Pravda*, Stalin breathed fire and fury. Such praise would have been welcome during the war, but now Stalin insisted that it was simply a cover for Churchill's hostile intentions and that Molotov should have recognized it as such.

Soon, it was reported in the British press that, according to high-level sources in Moscow, Stalin's power was not as great as many outsiders believed and government affairs were perfectly capable of being carried on without him. Incensed, Stalin lashed out at his designated heir, who, even if he were not the actual source of such statements, should have undertaken to suppress them. Stalin set his other satellites, Georgi Malenkov, Lavrenti Beria, and Anastas Mikoyan, against Molotov. They vied to denounce him for, among other outrages, consenting to an interview with the journalist Randolph Churchill. ("The appointment with Churchill's son was canceled because we spoke against it.") At length Molotov managed to stay afloat by tearfully admitting his mistakes to his rivals and penning a cringing letter to Stalin. Molotov kept his job, but from then on Stalin refrained from speaking of him as his successor.

Thus Stalin had put Molotov and the others on notice that he was always watching and that they ought not to grow too lax or too ambitious. Now, he had to dispel the rumors and to leave the world in no doubt that he, Stalin, was still number one and that he meant to keep things that way. By insisting to the Soviet people that their wartime allies in the West had already become their postwar adversaries, Stalin set himself up as the warrior whose duty it would be to drive back the enemy and save the Communist motherland—again. Under the circumstances, he simply could not contemplate retirement.

In this speech, Stalin bore no resemblance to the man Roosevelt had mistaken him for. Truman was beginning to recognize the need for a new approach to Soviet relations, one based on facts rather than on wishful thinking. Analysis of Stalin's presentation having yet to arrive from the US embassy in Moscow, Churchill's take on what was going on at the Kremlin was suddenly of particular interest. And the visitor had something even more important to offer. At a time when Truman had yet to emerge from Roosevelt's shadow, it might be difficult politically to depart from his predecessor's Soviet policy. The Fulton speech, delivered by a private citizen who also happened to be a master of the spoken word, as well as a figure of exceptional appeal to Americans, would allow Truman, at no political cost to himself, to see if the public was ready to accept a change.

After he met with Truman, Churchill spent the night at the British embassy. He had planned to return to Florida the next day, but snowbound airfields caused him to stay an additional night. Besides, the difficult trip north had left him feeling bilious and unsteady on his feet. His condition persisted in the days that followed. Back in Miami Beach, he remained in bed when he received James Byrnes, the US secretary of state, for two hours of talks. Following Churchill's White House visit, an announcement had gone out that he and Truman would fly to Missouri together on March 4. In view of his health, it was later quietly agreed that they would travel by train instead. Less than forty-eight hours before he and Clementine Churchill left for Washington on the first leg of his trip, he was coughing and complained on the phone to a friend that he was unwell.

Again, the timing of Churchill's appearance in the capital was fortunate. Again, actions taken by Stalin the day before Churchill arrived gave point to the visitor's argument. On March 2, the Churchills were en route from Florida when Stalin failed to heed the deadline by which it had long been agreed that all Red Army troops would be withdrawn from Iran. It was the first flagrant violation of a treaty obligation since Hitler, and commentators in the US and Britain were soon anxiously comparing it to the Führer's march into the demilitarized Rhineland in violation of the Treaty of Versailles a decade before. The Rhineland episode had been only the first of many such unilateral violations. Would Iran prove to be the same?

In the fortnight since Churchill's visit, the State Department had received an eye-opening message from the US embassy in Moscow. US chargé d'affaires George Kennan had long been frustrated by his government's naive view of Stalin. He used the present opportunity to put his considerable literary skills to work limning the postwar Soviet mind-set. Widely distributed and much read within the administration, the 8,000-word cable known as "the Long Telegram" did much to alter attitudes left over from the Roosevelt era. It fell to Churchill, however, to test the waters publicly. When Truman reviewed the final draft of the Fulton speech as they traveled on the ten-car presidential special to Missouri on the 5th, he called it admirable, said it would do nothing but good, and predicted it would cause a stir. Nevertheless, Churchill understood from the outset that he could count on White House support only if his presentation was well received. If he sparked off a controversy, he was on his own.

Resplendent in red robes that prompted some spectators to remark that he resembled a well-fed cardinal, Churchill made his case about time, the bomb, and the Soviet menace to an audience of 2,600 in the college gymnasium. Billed as the opinions of a private individual with no official mission or status of any kind, his comments were broadcast on radio across the US and reported around the world. As he talked on, he alternated between holding the chubby fingers of his left hand splayed across his round torso and using that hand to drive home a point.

He spoke again (though most listeners would be hearing that arresting phrase for the first time) of an "iron curtain" that had descended across the continent from Stettin in the Baltic to Trieste in the Adriatic. As he had done with Stalin at Potsdam, he ticked off the names of the eastern and central European capitals now under Soviet control. "This is certainly not the liberated Europe we fought to build up. Nor is it one which contains the essentials of permanent peace." He rejected the idea that another world war was either imminent or inevitable, and he argued that Soviet Russia did not at present desire war, but rather "the fruits of war and the indefinite expansion of their power and doctrines."

Churchill observed that in his experience, there was nothing the Soviets respected so much as strength and nothing for which they had less respect than weakness, particularly military weakness. He called on Britain and the US to emphasize their "special relationship," in the interest of being able to negotiate from a position of strength. The danger posed by Soviet expansionism would not be removed by closing one's eyes to it; it would not be removed by waiting to see what happened or by a policy of appeasement. What was needed was a settlement. The longer a settlement was put off, the more difficult it would be to achieve and the greater the danger would become.

Returning to the theme of fleeting time which he had sounded in his address in the House of Commons on August 16, 1945, he emphasized the necessity of acting in the breathing space provided by one side's exclusive possession of the atomic bomb. "Beware, I say; time is plenty short. Do not let us take the course of allowing events to drift along until it is too late."

The Fulton speech set off an avalanche of criticism and controversy in the US. In the wake of Stalin's remarks the previous month and of the Red Army's failure to leave Iran, there was perhaps little room to quarrel with Churchill's blunt review of the unpleasant facts. His recommendations were another matter.

Members of Congress lined up to administer a vigorous spanking to Churchill for—as they had heard him, anyway—proposing an Anglo-American military alliance, calling on Washington to underwrite British imperialism, and nudging the US in the direction of a new war. At the time, Halifax privately compared Churchill's situation in the US to that of a dentist who has proposed to extract a tooth. His many detractors were not so much claiming that the tooth was fine (given the recent news, how could they?), only that the dentist was "notorious for his love of drastic remedies" and that surely modern medicine offered "more painless methods of cure."

When he spoke in Missouri, Churchill had been careful to call attention to Truman's presence on the same platform and to point out that the president had traveled a thousand miles "to dignify and magnify" the occasion. Truman had applauded Churchill's address for all to see and he had praised it to him afterward in private conversation. In view of the uproar, however, he was quick to distance himself publicly. Three days after Fulton, he claimed not to have read the speech beforehand, and he declined to comment now that he had heard it. He wrote to his mother that while he believed the speech would do some good he was not ready to endorse it yet. Other figures associated with the administration also ostentatiously backed off. Secretary of State Byrnes denied advance knowledge of the content of the speech, though he, like Truman, had been shown a copy by Churchill himself. Under-Secretary of State Dean Acheson abruptly canceled a joint appearance with Churchill in New York.

In the belief that his views had been misrepresented in Congress and in a broad swath of the American press, Churchill spent the next two weeks trying to undo some of the damage. He made widely reported speeches and public appearances, but he also did some of his most important and effective work behind the scenes in Washington and New York. In one-on-one sessions with journalists, government officials, military leaders, and other opinion-makers, he patiently and methodically pointed out that he had called for a fraternal association, not a military alliance or a treaty. He maintained that, contrary to popular fears, he did not expect the US to back British foreign policy in every respect, or vice versa. He clarified that he had asked for a buildup of strength in pursuit of negotiations and that his purpose, as laid out in the text of the speech, was to prevent another war, not to start one. Throughout, Churchill toned down his language considerably; it was not his natural idiom perhaps, but it was what he felt people wanted to hear.

Halifax judged that by the time Churchill finished meeting with everyone on his long list he had made himself "a far more popular figure" than he would have been had he returned to England immediately after Fulton. And he had done much to put across his argument that Soviet expansionism was a topic the US was going to find it impossible to evade. All in all, Churchill had provided, in Halifax's view, "the sharpest jolt to American thinking since the end of the war."

He also produced a jolt in Moscow, though the Soviets waited several days to speak out. Churchill was in Washington preparing to go on to New York when the news broke that *Pravda* had run a front-page editorial headlined "Churchill rattles the saber." The piece denounced him for calling for an Anglo-American military alliance directed against the Soviet Union. A similar assault ran in the newspaper *Izvestiya* the following day. The day after that, Moscow radio broadcast a blistering attack by Stalin himself.

Speaking to an interviewer, Stalin called Churchill a "warmonger," compared him to Hitler, and accused him of seeking to assemble a military expedition against eastern Europe. He seized on Churchill's address as an opportunity to put a face on the danger from the West which he had evoked in his speech of February 9 to the Soviet people. George Kennan characterized Stalin's comments as "the most violent Soviet reaction I can recall to any foreign statement." In a curious way, Churchill had actually done Stalin a favor. The potential aggressor that Stalin had set himself up to defeat need no longer be an abstraction—Churchill was the threat personified. As Molotov later said, the Fulton speech made it impossible for Stalin to retire.

Stalin in turn gave Churchill a boost when he attacked him. Bypassing the elected leaders of Britain and the US, Stalin portrayed the emerging East-West conflict as a personal contest between Churchill and himself. At a moment when the news of Soviet troop movements in Iran and of US protests to Moscow over its actions not only there but also in Manchuria and Bulgaria were heightening public fears about Soviet intentions, Stalin drew Churchill into a debate that conferred upon him the unique status of the voice of the West. When Stalin pounced on what were after all the remarks of a private citizen, he ratcheted up the drama as Churchill could never have done alone.

On March 14, after a stack of evening newspapers with articles about the Stalin interview had been delivered to Churchill's twenty-eighth-floor suite at the Waldorf Towers, he sent word to reporters in the lobby that he would make no statement—yet. He was, however, set to speak at a banquet in his honor the following night in the hotel's grand ballroom, and he let it be known that he believed his comments would be of world interest.

Friday, March 15, proved to be foggy, rainy, and windy. In spite of the downpour, Churchill insisted on sitting on top of the back seat of an open touring car at the head of a twelve-vehicle motorcade which advanced at a walking pace. On both sides, a row of raincoated policemen flanked the car, provided by the city of New York, which flew an American flag above one headlight and a British flag above the other. The rain flattened Churchill's few remaining wisps of ginger-gray hair and streamed down his snub nose and jutting lower lip. Confetti clung to his blue overcoat as he held up a soggy black homburg to New York.

That evening, double rows of as many as a thousand demonstrators, dubbed "Stalin's faithful" by the local press, formed outside Churchill's hotel two hours before the banquet. Protestors carried picket signs, chanted, "GI Joe is home to stay, Winnie, Winnie, go away," and distributed reprints of a Communist *Daily Worker* cover showing a military cemetery with the headline, "Churchill wants your son." Mounted police maintained order, especially near the revolving doors where invited guests, including the mayor, the governor, and numerous ambassadors and other diplomats, were to enter. (The Soviet ambassador, notably, had sent last-minute regrets.) Inside, police detectives dressed in evening attire guarded the grand ballroom where four orchid- and carnation-laden daises had been set up in tiers onstage. A tangle of microphones marked the spot where it was widely expected that Churchill would reply to Stalin.

As the hour of Churchill's talk drew near, Manhattan bars and restaurants filled with people eager to hear him. One midtown restaurant promptly lost much of its business when its radio failed to work at half past ten. The proprietor of another East Side spot marveled that he could not recall a broadcast listened to by so many people or with such avidity since late 1941. Churchill came on the air twelve minutes later than scheduled, and the ovation he received at the Waldorf kept him from starting for an additional minute. At last, the familiar dogged, defiant voice on the radio answered Stalin's challenge to the Fulton speech by saying, "I do not wish to withdraw or modify a single word."

Churchill was back at the center of great events, where he loved to be, but the exertions required to get there had cost him dearly. On the night of the broadcast he was in splendid form, but in the days that followed he experienced dizzy spells. Once or twice, as he rose from a sitting position he began to fall forward and had to steady himself by grabbing his chair. He later said that acting as a private individual rather than a prime minister had been like "fighting a battle in a shirt after being accustomed to a tank."

VII

IMPERIOUS CAESAR

Southampton, England, 1946

A fur coat draped over his bowed shoulders, Churchill waited in the disembarkation shed at Southampton for his car to be brought round. During his nine weeks abroad, the political landscape at home had altered subtly but significantly. It was a measure of how much had changed that, two days before, when Stalin announced plans to withdraw from Iran he had felt the need to tell the world that his decision had not been prompted by anything Churchill had said in America.

On the other hand, much in political London remained the same. The Edenites were hoping to oust Churchill; Macmillan and Butler, perceiving elements of dissatisfaction with the interim Tory leadership, were jockeying to undermine Eden; and Eden himself was intent that that night, March 26, 1946, was the night when he would finally (in Cranborne's words) "grasp the nettle" and make a forceful case to Churchill about why it would be best if he retired.

At a moment when Churchill had begun again to feel his power, he was coming home greatly alarmed by how physically weak he felt. The dizzy spells had persisted, and there was concern that they could be the precursors of a stroke. At the Southampton quayside, he deflected questions about when he would next appear in the House of Commons by saying that he did not yet know the state of business in the House. He would have more information as soon as he had dined with his deputy.

A soupy fog in the English Channel had caused Churchill's ship to dock two hours late, so Eden was already waiting for him at Hyde Park Gate. On various prior occasions Eden had struggled to suggest that Churchill stand down in his favor. At the last minute, something had always caused him to hesitate. This time, he was confident things would be different—not because of any change in himself, but because Churchill's circumstances had changed. Initially Eden had taken a cynical view of the Fulton speech. He had remarked in private that he feared Churchill might actually be willing to set off another war in the hope of regaining the premiership. In the three weeks since Fulton, however, Eden had begun to sense that all the attention Churchill had been getting of late could prove useful to those who wished to force him out as Tory leader. In the past, Churchill had resisted any suggestion that he abandon power. But given his egotism and love of the limelight, might he not now be inclined to concentrate on his headline-making Soviet crusade and leave the conduct of party affairs to Eden?

Despite the long wait, Eden was in a hopeful mood when Churchill arrived at nine, but his plans quickly went awry. Before Eden could bring up the subject of Churchill's retirement, Churchill caught him by surprise. He, too, had a proposal to make this evening. Concerned about his waning strength, Churchill had devised a plan to allow him to hold on to the Conservative leadership without overtaxing himself. Just when Eden was about to ask the old man to step aside, Churchill asked Eden to help make it possible for him to keep his job. Churchill wanted Eden to take over for him officially in the House of Commons, as well as to assume the day-to-day work of running the party, while Churchill retained the overall leadership. As he was aware that Eden was financially pressed, he had already asked James Stuart, the chief whip, to see if a way might not be found to pay Churchill's salary as opposition leader to Eden instead. He went on to assure Eden that the arrangement was temporary, that he intended to keep the leadership for just a year or two and that his successor would benefit from having an opportunity to establish himself.

Few things could have been more insulting to Eden than the suggestion that he still had anything to prove, and few could have been more exasperating than the implication, heard so many times before, that he need wait only a bit longer before the prize was his. Again, the details of the handover were hazy. Again, Churchill set no firm date for his departure.

There was resentment on Churchill's side as well. An old man does not like to feel that he is being watched by "hungry eyes." When at some point in the discussion Eden managed to suggest that Churchill give up the leadership altogether, Churchill refused. And Eden, though he did not reject Churchill's offer in so many words, did not accept it either. The encounter on which Eden had pinned his hopes ended in bitter stalemate.

Having informed the press that he had no idea when he would next visit the House of Commons, Churchill disregarded his state of exhaustion and made a strategic surprise appearance on the opposition front bench the next day. Entering to his usual ovation, he let it be known that he intended to make his first speech in April during the budget debate. Eden, whose deputy leadership was widely deemed to have been a success, shrank to a subordinate position beside Churchill.

Moran arranged for his patient to be examined by the neurologist Sir Russell Brain, who concluded that the dizzy spells were nothing to worry about, that Churchill had merely overstrained himself in America, and that the episodes would soon pass. Thus reassured, Churchill seemed to forget his worst health worries and began to recover. He did not, however, forget Eden's bid to unseat him. When, over the vehement objections of his wife, Churchill took on the party leadership in 1940 in addition to his duties as wartime prime minister, it had been in part to keep the job from going to a younger rival who might later pose a threat to his premiership. In a similar vein, when he anointed Eden during the war he had been blocking the emergence of a more potent rival, someone less reluctant to seize the crown. In that sense, Eden's designation as heir apparent had been far from a sign of approbation.

Eden was still stoutly insisting to supporters that he would never accept Churchill's offer of the opposition leadership in the House without the party leadership overall when Churchill tripped him up by abruptly withdrawing it. Suddenly it was no longer in Eden's power to accept or refuse. Churchill indicated that as he was already feeling better, a formal arrangement was no longer necessary. Eden would still be called on "in an ever-increasing measure" to fill in for him in the House, but without any official status or salary. Churchill now expected Eden to do it all for nothing. The object of this division of powers was no longer to conserve an ailing man's strength; it was to spare Churchill what he saw as the drudgery of routine party business. In essence, Churchill wanted to do the work he chose, when he chose to do it. He wanted to speak and act when the spirit moved him—and to dump the rest of the job on Eden.

This time there was no display of temper on Churchill's side. On the contrary, in his note to lay out the new terms he addressed Eden with ironic courtesy, assuring him, even as he joyously twisted the screws, that he looked forward to working together "in all the old confidence and intimacy which has marked our march through the years of storm."

Still, Churchill made it clear that this was an offer Eden could not refuse—if, that is, he hoped to retain his claim to the succession. As if Churchill were innocent of Eden's nightmare of being overtaken by other claimants, he went on enthusiastically to propose Macmillan ("certainly one of our brightest rising lights") as a candidate to become the next party chairman. There was probably only one other name Churchill might have mentioned that would have been as likely to cause Eden to gag. Reminded that he was dispensable, Eden backed away from his demands. Eden timidly assured Churchill that he could count on him "to play my part."

Cranborne was horrified. He had spent the past few weeks in Portugal for his always precarious health, but he had been avidly monitoring all the moves and counter-moves from afar. He worried that under Churchill's leadership the postwar Conservative Party was fast becoming a kind of dictatorship. Cranborne fully shared Churchill's anxiety about the Soviet threat in Europe. Nevertheless, he was appalled that Churchill had delivered the Fulton speech without bothering to consult his Conservative colleagues beforehand. In the process, Churchill had committed what Cranborne saw as a political blunder which could have been avoided had Churchill taken the trouble to listen to other views. To date, Cranborne had been pleased to see the Labour foreign secretary consistently stand firm against the Soviets. On this matter at least, the Conservatives had been in the position of being able to sit back and support Bevin when necessary. Bevin had had to endure a good deal of sniping from the left wing of his own party, which remained infatuated with Moscow, but the broad unity of the country had been maintained. To Cranborne's eye, Churchill had unwisely destroyed that "happy unity": thanks to Churchill, opposition to the Soviet Union had become the policy of the party of the right and not of Bevin, whose position with his own supporters had thereby been made vastly more difficult.

Apart from all this, Cranborne believed there was a larger issue at stake. In important respects, Churchill was a lone wolf who disdained the pack. Cranborne regarded the Fulton speech as typical of Churchill's lifelong tendency to act without concern for his colleagues' opinions or his party's best interests. As far as Cranborne was concerned, this was the sort of high-handed, self-serving behavior he and Churchill's legion of other critics had long fervently complained of. Churchill for his part shrugged off such criticism. In the present instance, he saw it as a matter of perspective: why concern himself with relative trifles like party interests or colleagues' wounded feelings when he was trying to head off another world war?

Thus the battle lines were drawn. Cranborne viewed Churchill as "imperious Caesar" who simply had to be stopped. If Eden lacked the will to force the issue of Churchill's retirement, it seemed to Cranborne that others were going to have to do it for him. Within days of his return to London, Cranborne was discreetly proposing that party leaders "take their courage in both hands" and make a joint approach to Churchill. He acknowledged that Churchill would probably never forgive them and that they might very naturally hesitate to participate. But, Cranborne stressed, he saw no alternative. When Eden discouraged him, Cranborne wrote in disgust to his father, Lord Salisbury, that he had been ready to lead a cabal against Churchill but that there was no reason to go forward as long as Eden refused to act.

Some of Churchill's longtime friends were also quietly advising him to retire, but their motives were very different from those of the Edenites. There was feeling among some of Churchill's contemporaries, such as the seventy-six-year-old South African prime minister Jan Smuts and the seventy-one-year-old Canadian prime minister Mackenzie King, that by allowing himself to be caught up in party strife he was tarnishing his reputation. An incident in the House of Commons on May 24 was a case in point. Churchill had caused a furor when, during a particularly fierce dispute, he stuck out his tongue at Bevin.

In contrast to those who wished to give Churchill the hook for personal or party ends, Smuts and King, who were in London for a meeting of Dominion leaders, were concerned solely with what was best for him. Churchill was especially fond of the South African leader, of whom he once said, "Smuts and I are like two old lovebirds molting together on a perch but still able to peck." On the present occasion, Smuts advised Churchill to retire immediately.

In a similar vein, Mackenzie King, who was then beginning his twentieth year in office, recommended that Churchill remove himself from the hurly-burly of domestic politics in favor of taking a larger view in keeping with his titanic stature. Both he and Churchill were now the very age Lord Fisher, the former First Sea Lord, had been in 1911, when, Churchill recalled in *The World Crisis*, "I was apprehensive of his age. I could not feel complete confidence in the poise of the mind at 71." As Churchill well knew,

King had begun to worry about his own fading powers. As a consequence of the uproar over Churchill having stuck out his tongue at Bevin, claims had been heard from the Labour benches that Churchill had entered his "second childhood." Might the time have come to bow out for dignity's sake? Churchill was firm that it had not.

He made his intentions clear at a dinner party on June 7 in honor of Mackenzie King's long service, hosted by Clement Attlee in the paneled dining room at 10 Downing Street. Field Marshal Smuts was present as well. When the conversation among the old men turned to Roosevelt's state of mental and physical deterioration at Yalta, Churchill suggested that he could close his eyes and see the ruined president as he was then. Someone chimed in to speak of Gladstone, who had been returned to power at a great age. In response, Churchill expressed confidence that he had time yet. But did he really mean to suggest that he believed he could be prime minister again? When King urged him to devote himself to authorship rather than politics, Churchill shot back that he had no intention of abandoning the fight and planned to lead his party to victory at the next election. In a separate conversation, he told Betty Cranborne (who later repeated it her husband) that nothing would induce him to retire.

He dropped his bombshell to Eden when the latter returned from a three-week trip abroad. Mindful of the work that faced him in preparing his memoirs, Churchill suggested that he might be willing to renew the offer of officially dividing the Tory leadership and of transferring his salary to Eden. This new offer would differ from the previous one in a crucial respect. In March, Churchill had assured Eden that he meant to keep the leadership for a limited time. Three months later, he told him what he had already told Mackenzie King and others about his determination to recapture the premiership.

It took Eden a while to absorb the astonishing news. Churchill, after all, had gone from pledging to retire at the end of the war, to promising to stay on as party leader for no more than two years—to this. After the Conservative rout, Eden's sole consolation had been that when next the tide of Toryism came in, Churchill could not possibly still be in position. Now, Churchill was confidently suggesting it was possible.

In the tortured weeks that followed, Eden by turns doubted that Churchill could be right about his prospects, wondered whether in light of Churchill's comments he had better give up politics in favor of a career in finance, told himself and others that Churchill was likely to take a more realistic view by the end of the summer, and strongly considered trying to find a way to accept Churchill's offer—if, that is, he ever actually made it.

Cranborne suggested to Eden that, under the circumstances, it might be best at this point simply to stand down as second-in-command and take his own independent line in Parliament. He was urging Eden, in effect, to abandon the security of his role as designated heir and to fight for the crown alongside any other contenders. The proposal reflected the considerable freedom of action Cranborne's position as heir to the Marquess of Salisbury conferred. He cared a good deal less about office and security than Eden, but then he had the luxury to be inflexible and to put practical considerations aside. In 1938, when Eden and Cranborne resigned as foreign secretary and under secretary respectively, some observers who knew both men believed that Eden had bailed out only because he had been pushed (or was it shamed?) by Cranborne. Eden's resignation speech in the House of Commons had been, to some tastes, disappointingly soft and vague in contrast to his friend's forthright remarks. Cranborne had bluntly accused the prime minister of surrendering to Italian blackmail. (Chamberlain said of Cranborne: "Beware of rampant idealists. All Cecils are that.") Hoping to protect his claim to succeed Chamberlain, Eden had been careful not to burn his boats irretrievably with the party. In any case, Eden would long be distressed by the perception that he had been—indeed, still was—in thrall to Cranborne's more powerful personality.

What made this all so painful was that there was much truth

to the picture. To Eden's simmering frustration, with Cranborne, as with Churchill, he was and perhaps always would be number two. On the kingmaker's side there was friendship and loyalty, to be sure, but there was also a tendency that did not go unnoticed in their inbred aristocratic world to treat Eden "rather as if he were the head butler at Hatfield House" (the Cecil family seat).

Eden, meanwhile, continued to vacillate, and Churchill went off to Switzerland without having made another concrete offer to share power. Churchill drew more world headlines when he spoke at the University of Zurich on September 19. His remarks were the second installment of his prescription for confronting the Soviet danger. Having already called for an Anglo-American partnership to counter the massive Soviet presence in the occupied territories of Europe, he now proposed an end to retribution against vanquished Germany. He declared that Germany must be rebuilt and he argued that France must lead the effort. Churchill urged listeners to turn their backs on the horrors of the recent past and to look to the future—in other words, to welcome the Germans into the community of nations. So soon after the war, his recommendations were strong medicine, but, as he admitted privately, he saw a rebuilt Germany as a necessary defense against the Soviet Union. It was Churchill's hope that the creation of a strong Europe led by a revitalized France and Germany would do much to avert a war with the Sovict Union, and to produce a lasting settlement at the conference table.

In reaction to Churchill's call for the rebuilding of Germany and the formation of a United States of Europe, Moscow radio accused him of seeking to unite the continent in preparation for war. When Churchill went on to ask publicly why the Soviets were maintaining so many troops on a war footing in the occupied territories of Europe, Stalin called him the worst threat to peace in Europe.

Churchill was already under heavy fire from Moscow when he turned up in Paris to discuss the situation in Europe with US Secretary of State Byrnes, who was there for the peace conference. Frustrated by his inability to extract the information from his own government, Churchill was eager to be brought up to date on current thinking in Washington. He also wanted to maintain his personal contacts with the Americans. Bevin, for his part, could not see why. Was Churchill's party not out of power? What business did he have in Paris? To the acute irritation both of the Labour government and of certain of his Conservative colleagues, Churchill seemed to be running some sort of high-flying, out-of-control, one-man foreign policy shop. Official negotiations with Molotov continued to drag, and the British foreign secretary was furious at the prospect of Churchill, who had not been invited to participate, doing or saying anything in the course of his short stay to complicate matters.

In anticipation of Churchill's arrival in Paris, there had been much agitated discussion within the British delegation about how best to cope. Bevin worried that allowing Churchill to stay at the embassy would seem to confer government approval on his private talks with Byrnes and other officials, when in fact Britain had no control over anything he said. Duff Cooper, the British ambassador, successfully argued for accommodating him there, the better to manage him. Afterward, the ambassador wrote of his fellow Conservative's whirlwind visit with a mixture of amusement and annoyance (clearly more of the former than the latter), "Having possibly endangered international relations and having certainly caused immense inconvenience to a large number of people, he seemed thoroughly to enjoy himself, was with difficulty induced to go to bed soon after midnight and left at 10 a.m. the next morning in high spirits."

A year after Churchill returned from his Italian holiday, he had reason to be high-spirited. Though out of office, he had handily regained influence. Though Britain had a new government, he regularly managed to upstage it. Whether or not one sympathized with his arguments in Fulton and Zurich, there could be no denying that he had framed the international debate on such matters as Soviet expansionism and European reconstruction and unification. At the first annual Conservative Party conference since the war, held in Blackpool in October 1946, Churchill avowed that while it would be easy "to retire gracefully," the situation in Europe was so serious and what might be to come so grave that it was his "duty" to carry on. As he approached his seventy-second birthday, he spoke with assurance of turning out the socialists and he remained confident of his ability to secure the peace if only he could get back to the table with Stalin.

Still, there was growing dissatisfaction in Conservative quarters with a leader who was absent much of the time, traveling, speaking, writing, and collecting awards. A fresh round of defeats in the December by-elections intensified Conservatives' hunger for a leader willing to devote his energies to remaking the party. It was a measure of how much the tide had begun to turn against Churchill that the Conservative chief whip James Stuart went to Cranborne to discuss the need for a change of leadership. Though at the end of 1946 Stuart remained distinctly unimpressed by Eden, he had sadly concluded that Churchill's spotty attendance in the House of Commons was making the conduct of business almost impossible. In conversation with Cranborne, Stuart proposed to speak to Churchill. He wanted him to revive his plan to hand over the opposition leadership in the Commons to Eden while retaining the broader leadership of the party. At least that way, someone other than Churchill would have real authority to lead in the House. Stuart judged that Churchill might be more amenable to sharing power now that he was so effectively influencing international opinion. Cranborne was a good deal less optimistic about what amounted to a first approach to Churchill by his colleagues, on Eden's behalf. Nevertheless, he gave the mission his blessing. It was a mission that few would have taken on willingly, but Stuart, a raffish Scot, had a reputation for fearlessness. The very fact that someone unaffiliated with the Eden faction was prepared to make the proposal might signal to Churchill that it was indeed time to think about going.

To Stuart's relief, Churchill responded calmly to the suggestion that he had already done so much for his country that he could retire and enjoy the rest of his life without regrets. Still, when Stuart proposed that for the good of the party Churchill consider reviving his plan to share the leadership with Eden, Churchill would not hear of it. Churchill explained that great events were pending, though not immediately, and that he wanted to be in a position to handle them himself. His answer went to the heart of what power meant to Churchill. Through the years, he had often suggested that office and title meant nothing to him; what appealed to him was the opportunity to direct events and to shape the future. And so, he made it clear to Stuart, it was now.

On a lighter note, Churchill addressed his colleagues' concerns about whether he was still up to the burdens of the opposition leadership by informing Stuart that he meant to install a bed in his room at the House of Commons. He assured the vastly amused chief whip that this would allow him to take naps there and no one need worry that he would be too tired to attend. Churchill insisted that Eden could wait a little longer to enter his inheritance and that Eden knew he was devoted to him. At the end of the hour, Stuart, veering between laughter and tears of frustration, had gotten absolutely nowhere. The most he could say was that at least the old lion had not bitten off his head.

For Cranborne, the news that Stuart had failed was unwelcome but not unexpected. By the end of 1946, he had explored what seemed like every option: he had prodded Eden to approach Churchill on his own; he had volunteered to organize a cabal; he had suggested to Eden that he abandon his role as designated heir and fight for power in the House; he had given the nod to the chief whip to act on Eden's behalf. Nothing had worked.

Cranborne lamented that Eden was "rapidly losing ground." He reckoned that the only way Eden could reestablish himself was "by some resolute step, such as he took when he resigned in 1938. That got him the reputation of a strong man, but he cannot live on this one incident in his career forever." It was a disturbing assessment of the man Cranborne still hoped to make the next prime minister of Great Britain. Cranborne insisted that if Eden wanted to be perceived as a leader he had better begin to act like one; until Eden made his move, there was nothing anyone else could do for him. In the interval, Churchill clearly meant to hold on to the party leadership "at all costs." For one thing, Cranborne reflected, Churchill liked power. For another, Churchill was convinced that "like Lord Chatham he can save England & no one else can." Cranborne did not intend the comparison to the aged, ailing eighteenth-century statesman who pushed himself to the limits of his physical endurance, collapsed on the floor of Parliament, and died soon afterward to be flattering. Nevertheless, Cranborne recognized that part of what made Churchill an especially formidable opponent in any attempt to challenge his leadership was that he really did think he was the one man to save his country.

Still, at that point there was no rational reason to believe that Churchill could ever be prime minister again. Labour remained overwhelmingly popular, and the wisdom continued to be that the Conservatives could not hope to recapture Number Ten for at least two five-year election cycles. For all that Churchill had accomplished since he left office, the arithmetic continued to be against him.

VIII

PLOTS AND PLOTTERS

Hyde Park Gate, 1947

The downstairs rooms of the house in Hyde Park Gate were dark and unbearably cold. It was Sunday, February 16, 1947, and ordinarily the Churchills would have been at Chartwell, but they had decided to stay in London on account of the heavy snow and freezing temperatures. Since the last week of January, Britain had been suffering the most brutal weather conditions anyone could recall. Government mismanagement of the recently nationalized mines had left the country without a sufficient supply of coal. The power system was at breaking point and heavy restrictions were in effect. Clementine Churchill had secured a doctor's certificate to allow her husband's bedroom to be minimally heated. Even so, as the use of electricity was prohibited from 9 a.m. until noon and again from 2 p.m. until 4 p.m. on pain of a heavy fine or imprisonment, she had also arranged to have Winston's bed moved near the window so he could work by natural light.

Churchill liked his comforts and one might have expected to find him in a petulant mood. On the contrary, the former prime minister, propped up against a mountain of pillows, his elbows resting on sponge pads on either side of his bed table, was sunshine itself. It is unattractive to gloat over other people's misfortunes, especially when one is poised to profit from them. But under the circumstances, who could blame him? After the Labour landslide he had argued that, like it or not, the opposition would have to wait upon events. Those events had come in force. The government was blaming its troubles on an act of God. For Churchill, the arctic weather was a godsend. Suddenly, it seemed as if Labour might indeed be vulnerable at the next general election.

The revelation of official ineptitude in miscalculating both the amount of coal needed to sustain Britain that winter and the productive capabilities of the mines shattered public confidence in the socialists. Popular disappointment was so enormous because the expectations of a better material life after the war had been unrealistically high. The continuous snowfall paralyzed the already feeble economy, and, rightly or not, many Britons blamed Number Ten for the stalled train lines, business closings, mass unemployment, food and water shortages, long sluggish queues, and overall discomfort and deprivation. They also blamed the government for their nation's abruptly diminished place in the world when, in the midst of the crisis, Britain made it clear that it could no longer afford to keep up its military commitments in Greece and Turkey.

In March the snow and ice gave way to pounding rains and catastrophic floods. Again the economic consequences were devastating, and again the government struggled to cope. Churchill asked the House of Commons for a vote of censure. The Conservatives were still vastly outnumbered and, as he knew it must, the vote on March 12 went against him, but this time anti-government sentiment was more pervasive than before. He had the support of Liberals and some independent MPs, and there was fierce disagreement among the Labour members about how best to respond to the crisis. The day after the vote in Parliament, both Macmillan and Butler confidently suggested to a large and enthusiastic Conservative meeting in London that a new general election could be in the offing sooner than anyone had thought. If the economy continued to deteriorate and the socialists persisted in fighting among themselves, the government might be brought down even before 1950.

Conceivably, the coal crisis had gained Churchill five years or more—no small gift to an old man. But the events that had caused him to smell Attlee's blood also made it seem more urgent than ever to his Tory adversaries to dislodge him lest he still be in place when a new election was called. It had been one thing for him to cling to his job when the party had no realistic chance of being returned to power. But everything had changed, and beginning in late February, there was a flurry of small private meetings of Conservatives, most but not all of them Edenites, anxious to see Churchill go.

When Churchill absented himself in June for five weeks after a long-postponed hernia operation that had been troubled by complications, his opponents thought they might have caught a whiff of his blood as well. After all, the announcement had lately been made of Churchill's deal to be paid more than a million dollars for the US book and serial rights to his war memoirs, the first volume of which was scheduled for publication in 1948. Researchers and other staff had been hired, permission to draw on his official wartime papers had with much difficulty been obtained, and work on the text was under way. Given the deadline and Churchill's always precarious finances, not to mention his health, might he not be amenable to a plea that it really would be best for everyone if he stepped aside sooner rather than later?

Eight senior Conservatives gathered in the upstairs drawing room of the Tory MP Harry Crookshank to select an emissary to make the case to Churchill. Crookshank, who had been castrated by a burst shell during the First World War, lived with his mother in Knightsbridge. Eden, notably, skipped the meeting on the grounds that his direct participation in a plot to install him would be awkward. But, though he had previously vowed to do no more until Eden acted, Bobbety Cranborne—who had become the 5th Marquess of Salisbury on the death of his father in April—was again a key player in the machinations. Why did Salisbury (as Cranborne was now known) stick with Eden when he perceived his flaws so clearly? Quite simply, in Salisbury's view there was no other viable candidate. He was opposed to Butler because of his history as an appeaser in the 1930s. Nor was he prepared to back Macmillan, whom he disliked and distrusted. Salisbury's wife, whose opinions meant much to him, was also no fan of Macmillan's. The reasons for Lady Salisbury's antipathy were strange and complicated. Early in her marriage she had lost interest in Bobbety and began to have affairs with other men. Her husband's rise to political prominence rekindled her interest, and she became fiercely possessive of him. She even resented his lifelong affection for his sister, Mary, Duchess of Devonshire, with whom Macmillan, who was married to the Duke's sister, enjoyed a close, confiding friendship as well. Macmillan's association with "Moucher" Devonshire doomed him in Betty Salisbury's unforgiving eyes.

Though Salisbury would never have admitted as much, there was another compelling reason to keep coming back to Eden. Salisbury could never have pushed around Butler or Macmillan the way he did Eden. They would not have tolerated it. The very weakness that Salisbury deplored in Eden in some respects made him a most attractive candidate in others. If it was influence Salisbury hankered for, Eden was assuredly his man.

There was unanimous agreement among Crookshank's guests that Churchill must go, but most were unwilling to face him. It was not just his epic temper that daunted them. If Churchill survived the putsch, the mission might be a career-destroyer for any ambitious Tory who consented to undertake it. Finally, James Stuart agreed to try—again.

Interestingly, Butler, who attended the Crookshank luncheon, also threw in his lot with a group of Labour members who, in the hope of saving their own hides should Attlee fall, aimed to bring down the government themselves in favor of a coalition headed by Ernest Bevin. Because of the hard line he had taken on the Soviets, Attlee's foreign secretary was perhaps the one Labour figure capable of commanding substantial Tory support. What was in it for Butler? As matters stood within the Conservative Party, his way to the top was blocked by a number of factors. One objection was that his résumé was too thin. Another, which threatened to be insurmountable, was that he had been too closely associated with the policy of appeasement when he served in the Chamberlain government. Then, quite simply, there was the perception that the succession had long been fixed. Were Churchill to go, Eden was ready, as well as widely expected, to take his place. Should Butler be part of a coalition that bypassed Churchill, neither Churchill nor his designated heir would any longer stand in Butler's way.

In July, Churchill returned from his convalescence to be greeted by applause from all parties—and by a whirligig of plots and plotters. He was sitting in his room at the House of Commons when Stuart came in. Stuart began by saying that he had a difficult task to perform and that he hoped Churchill would bear with him without being annoyed. When last they spoke of retirement, Churchill had reacted calmly, and Stuart hoped that might be the case again. He repeated what he had said previously about no other man having done more for his country than Churchill. Then he went on to report the view of their colleagues that the time had come for a change of leadership.

Churchill exploded, "Oh, you've joined those who want to get rid of me, have you?"

"I haven't in the least," Stuart protested, "but I suppose there is something to be said for the fact that change will have to take place sometime and you're not quite as young as you were."

Churchill responded by angrily banging the floor with his walking stick. With that emphatic gesture, he put an end to the cabal to unseat him in favor of Eden. The plotters acknowledged that Churchill was too beloved a figure, both within the Tory rank and file and the country at large, for there to be any public perception that party leaders had forced him out. If he went, it had at least to appear to be of his own accord. And he was unlikely to go anywhere at a moment when the premiership seemed achingly and unexpectedly within grasp.

Churchill was not supposed to have learned of the negotiations between Butler and other Tories, on the one side, and Attlee's chancellor of the exchequer Hugh Dalton, on the other, but there had been much open talk of them, including some indiscreet remarks by Butler's wife. Sydney Butler could not resist broadcasting the news to at least one lunch partner that Churchill was about to have a "rude awakening" when a coalition government was formed with someone other than himself at the top. In the end, however, it was Sydney's husband who was in for a jolt when Churchill appeared unannounced on July 31 at a meeting of the 1922 Committee, the official Conservative organization for backbenchers. Without referring to Butler or the others by name, he put the plotters on notice that he knew of their machinations. Speaking as if it were a foregone conclusion that were there to be a coalition he would be at the head of it, Churchill warned against any such arrangement with Labour on the grounds that it would deprive the country of an effective alternative government. He made many of his listeners' mouths water at the prospect of their party's imminent return to full power, if only they proceeded judiciously—under his leadership, of course. Why agree to share power, Churchill suggested, when Conservatives could have it all? Why indeed, many backbenchers concurred.

Having outmaneuvered the conspirators in his own party, Churchill went on to leave no doubt in the minds of Britons that he and no other man would lead the Conservatives to victory. At a time when the government was announcing stringent emergency measures that included longer working hours and extensive further rationing, Churchill publicly declared his intention to fight the next election on the matter of the economic crisis.

Eden, who had just turned fifty, alternated between threatening (again) to forsake politics altogether and making it plain to friends that even now he thought "of nothing except becoming leader of the Conservative Party." The aging "boy wonder" seemed to be of two minds about his personal life as well. One minute he was pining for Beatrice Eden to return now that her American lover had gone back to his wife, and the next he was chasing young women including Kathleen "Kick" Kennedy, the twenty-six-year-old daughter of the former US ambassador to Britain, and Churchill's twenty-six-year-old niece Clarissa Churchill. In private, Eden made no bones about the fact that he loathed Churchill, yet he cheerfully played his part at the Tories' annual conference, held in Brighton that October, where Churchill was set to challenge the government to call a new election.

Churchill liked to use a basket of bread on his arm to lure and coax his pet birds at Chartwell—geese, black swans, sheldrakes, white swans, and ducks—and, finally, to set them all fighting. He enjoyed watching his would-be successors in the Conservative Party battle over crumbs as well. Eden and Butler vied for the party's attention in speeches, to some three thousand delegates from every corner of the nation, about Conservative ideas and how they might be applied in a new government. But it was Macmillan, the new party president, whose remarks seemed to claim the greatest share of the spotlight at Brighton.

The adroit, confident performer who introduced Churchill on the third and final day bore little resemblance to the shy, often ineffectual, even buffoonish figure Macmillan had cut before the war, when he had had a reputation as "the most boring speaker in the House of Commons." By Macmillan's own reckoning, he had all but reached a dead end in his career when, in 1942, Churchill assigned him to North Africa as the British government's representative to the Allies in the Mediterranean. At first he feared he had been sent to languish in "political Siberia," but it rapidly became apparent that the position would be the making of him. He gained stature and confidence on the job and earned the respect and affection of colleagues like Dwight Eisenhower. By 1944, Lady Diana Cooper, who had gone to Algiers with her husband, Duff Cooper, was predicting something that would have been unthinkable just two years before. She prophesied that Macmillan was going to be prime minister someday.

Macmillan, who had thinning gray hair which he brushed straight back from a high forehead, drooping eyelids, and a bushy mustache arranged to conceal irregular teeth, came home in 1945 with great hopes. But it soon seemed as if the late-starting momentum of his career might have been fatally interrupted. He lost his seat in Parliament in the general election that swept the Conservatives from power. The setback prompted Macmillan to brood about how little he had accomplished in life to date. Eden at least had been foreign secretary; but despite being past fifty, Macmillan believed that he himself had only just begun. As it happened, a safe Tory seat fell vacant shortly thereafter. Macmillan fought a by-election and was back in Parliament soon enough, but the effect upon him of the general election was very severe. Having briefly been convinced that his political life was over, Macmillan felt as never before the need to propel himself forward while he had the chance. He emerged as a shrewd operator, always maneuvering, always watching. Some people thought him devious and unscrupulous, but as far as Macmillan was concerned he was only making up for lost time.

Though Harry Crookshank was his close friend and James Stuart his brother-in-law, Macmillan refused an invitation to the witches' Sabbath to plot Churchill's ouster. But was it, as he professed, solely (or even largely) devotion to Churchill that motivated him?

When Macmillan first began to attract praise for his achievements in North Africa, Eden had pricked up his ears—and they stayed up for years to come. The foreign secretary became wildly jealous of the Minister Resident, a situation that was not helped by Churchill's habit, then and later, of taunting his heir with references to Macmillan's rising fortunes. As a consequence, were the prize to fall into Eden's hands at this point, he was unlikely to do any favors for Macmillan. From the perspective of Macmillan's ambitions, it therefore seemed better that Churchill be allowed to carry on, giving Macmillan a chance to overtake Eden, precisely as Churchill himself had once done.

At a moment of much discontent among senior Tories about the leader's refusal to hand over before the next general election, Macmillan cast himself as the most outspoken supporter of Churchill's ambitions to return to Number Ten. The other "possibles" and their associates had no alternative but to cheer along with the Conservative rank and file when Macmillan pointedly, some claimed unctuously, introduced the closing speaker at the Brighton conference: "We need a Prime Minister—Mr. Churchill."

IX

BEFORE IT IS TOO LATE

Westminster Abbey, November 1947

Churchill tended to operate on his own timetable, and it often fell to Clementine (or to one of his daughters when they were "in attendance") to make sure that he was reasonably punctual. But on the morning of November 20, 1947, not even Clementine's best efforts prevented the couple from arriving at Westminster Abbey after the approximately 2,500 other guests at Princess Elizabeth's wedding were already seated. The organist had begun to play, the scarlet-and-white-clad choirboys had filed in, and Prince Philip and his best man, both in naval uniform, were in position at the altar steps.

Guests on either side of the nave stood up for Churchill as he was escorted to his place beside Jan Smuts and Mackenzie King in Clement Attlee's pew. Mrs. Churchill was seated separately with Mrs. Attlee. In the seconds before a roll of drums and a fanfare of trumpets announced the appearance of the bride in the west door of the abbey, the three old men chatted affectionately. Field Marshal Smuts, wearing an overcoat and high boots, sitting between Churchill and King, said in a parched voice, "We shall not see the likes of this again. We shall soon be passing away."

This of course was far from how Churchill preferred to view his own immediate future. At a moment when both Smuts and King recognized philosophically that the time was fast approaching when they must allow younger men to have a chance, Churchill burned to lead Britain again. He reckoned that Labour was likely to put off a new general election as long as possible, which meant that he might have to wait until the summer of 1950 before he had a chance to reclaim power.

Churchill's sense of frustration was particularly acute in advance of the fifth session of the Council of Foreign Ministers, scheduled to open in London on November 25. Germany was the main topic on the agenda, and Churchill was hardly alone in his concern that a failure to reach an agreement there could set off another war. Failure, at any rate, seemed assured. In the belief that prolonged chaos would spur the Germans to go Communist, the Soviets did not see it as in their interest to rebuild Germany physically and economically as the US, Britain, and France proposed to do.

By this point, the Kremlin had reacted to extensive new US initiatives to limit the spread of Soviet power in Europe—the Truman Doctrine and, in particular, the Marshall Plan to rebuild Europe, for both of which Churchill could lay claim to having been an inspiration—by dropping any remaining pretense that the wartime alliance remained intact. The previous month, Moscow had decreed that the postwar world was divided into two camps, one led by the Soviets and the other by the Americans. Smuts and Mackenzie King were among the Dominion leaders who, during their visit for the royal wedding, received secret briefings at Number Ten which conveyed the Attlee regime's intense pessimism about the situation in Europe. Bevin, set to face Molotov at the foreign ministers' talks, told the Dominion leaders that, in the event the Soviets moved to block the Western powers' access to Berlin, fighting could break out as early as the following month. Nerves were on edge as a confrontation over Berlin threatened to spark off a third world war.

As far as Churchill was concerned, Germany was among the critical issues that should have been resolved by the leaders at Potsdam and ought never to have been passed on to foreign ministers. In his analysis, Soviet Russia, like Hitler's Germany at the time of the Rhineland episode in 1936, was not ready to fight another war.

He believed that this was the moment to call Stalin's bluff and use America's exclusive possession of the bomb to force the Soviets to release their grip on their satellite states in Europe and retire to their own country. Based on his assessment that the Soviets feared the bomb too much to face a war, he predicted that they could still be made to retreat without bloodshed.

It pained Churchill to be out of office while others undertook crucial negotiations he was convinced he could handle much more effectively himself. He believed that, faced with anyone else at the conference table, the Soviets might yet be inclined to seek further appeasement in Germany or elsewhere. "If I were there," Churchill told Violet Bonham Carter, "they would know very well what to expect from me."

On the day the foreign ministers' talks began, Churchill did most of the talking at a small farewell lunch party for Smuts and King at the home of Churchill's former chancellor of the exchequer John Anderson and his wife, Ava. Also present were Clementine Churchill and Harold Macmillan. Churchill argued that if their side failed to make it clear that the Soviet presence in Europe was unacceptable, another world war was a certainty, either directly or in a few years' time: "I am telling you now that I see the coming war as plainly as I did the last one." As Churchill held forth at the Andersons' table, his fervor and frustration were such that King, seated to his left, could see the greater part of the whites of his bulging eyes, which looked as if they were about to pop out of his head. "The gleam in his eyes was like fire," the Canadian prime minister wrote in his diary afterward. "There was something in his whole appearance and delivery which gave me the impression of a sort of volcano at work in his brain."

The foreign ministers were still closeted at Lancaster House when, on December 10, Churchill flew to Paris en route to Marrakesh. There he meant to "bulldoze" a finished manuscript, or something very close to that, of volume one of his war memoirs, which portrayed his role in the 1930s as a truth-teller and a prophet and ended with him becoming wartime prime minister. When, previously, Churchill's colleagues had urged him to devote himself to his memoirs, it had been their way of trying to persuade him to fade from the scene. Churchill had come to view the project very differently. Far from being an activity to occupy his time in retirement, the preparation of his Second World War history became an integral part of his campaign to catapult himself back into office.

As a politician, Churchill was gifted with the ability to reach people powerfully and directly in his capacities as an orator and an author. There could be no denying that the Fulton speech had done much to transform public opinion, both in the US and Britain. Now that people perceived the developing danger, Churchill used his war chronicles to help make the case that he was the leader to confront it.

After the party's defeat in 1945, senior Conservatives had been eager for Churchill to travel abroad and leave the management of party affairs to Eden. Much had altered since then, and Churchill's regular and prolonged absences had become a source of aggravation to some of the very people who had once hoped to see him go off for as long and as often as he liked. Bitter complaints were heard that Churchill was not in the least interested in politics; that he was preoccupied with his memoirs; and that he insisted on retaining the party leadership largely because he hoped to wipe out the disgrace of the general election. There was also annoyance at the prospect of his earning so much money from his books when he ought to have been looking after Conservative affairs in the House. There was upset that when he did make the occasional grand parliamentary appearance, he seemed to take an interest in no one's speeches or opinions but his own. And there was indignation that he failed to mingle with his party in the Smoking Room, only with the "admirers and hangers-on" who formed his claque.

These charges pumped new life into decades-old questions about Churchill's party loyalty; he had had a long and difficult association with the party he blamed for having forsaken his father. As a young man, Churchill had quit the Tories supposedly on a matter of principle, but some people insisted it had been to further his own personal interests. After he returned to the Tory party, he never entirely shook off the stigma of having crossed the aisle to the Liberal side. Now, fellow Conservatives asked again whether he was out to help his party or himself. The crusty veteran Tory MP Cuthbert Headlam complained that Churchill had "always been too much interested in himself to run a party."

Churchill's unabashed egocentricity did make him a less than satisfactory party leader, but, as even certain of his critics acknowledged, it gave him the strength and pertinacity that had allowed him to carry on in the face of apparently insurmountable obstacles. The conviction that he had been put on earth to do great things had sustained him during his wilderness years (when he had been violently at odds with his party on the matter of the German threat) and during his wartime premiership. It sustained him now.

By the time Churchill returned from his working holiday, the foreign ministers' talks had failed, precisely as he had predicted they would. When he rose to speak in the House of Commons on January 23, 1948, he asked the question that was on many people's minds: "Will there be war?" He reminded listeners of what he had said at Fulton, that he did not believe the Soviet government wanted war, but the fruits of war.

It would be idle to try to reason or argue with the Communists, he maintained. It was, however, possible to deal with them on a realistic basis. The best chance of avoiding a war was to bring matters to a head and to come to a settlement with the Soviet government "before it is too late." He recalled his speech on August 16, 1945, during the V-J Day celebrations, when he observed that America's exclusive possession of the bomb would provide a three- or four-year breathing space. He noted with alarm that more than two of those years had already passed.

He did not think that any serious discussion which it might be necessary to have with Moscow would be more likely to reach a favorable conclusion if their side waited for the Soviets to get the bomb as well. "You may be absolutely sure that the present situation cannot last."

The appearance that spring of the advance excerpts from volume one of Churchill's memoirs hammered home his case that steps must be taken at once to avert a new war. The *Daily Telegraph* in London published a total of forty-two extracts from Churchill's depiction of the run-up to the Second World War, the *New York Times* published thirty, and *Life* magazine serialized the book over the course of six weeks. The narrative spanned events between 1919 and 1940, but from first to last the story was framed in terms of what Britain's early mistakes with Hitler had to teach the world about present problems.

The timing of its publication caused the serial to resonate inescapably in terms of the Berlin crisis. Churchill noted in his introduction that after all the exertions and sacrifices that had been made in the last war, neither peace nor security had been attained, and mankind was at present in the grip of even worse perils than those it had just surmounted. He expressed the hope that "pondering upon the past may give guidance in days to come." The *Daily Telegraph* billed the work as an unchallengeable case against any repetition of the appeasement that had led to the Second World War. The *New York Times* cast it as a solemn warning that "unless we are able to learn from the past and avoid its mistakes, we stand in danger of seeing history repeat itself."

On the political front, the serial sparked outrage among Conservatives, who worried that it would jeopardize party prospects in the 1950 general election. When they had first advised their leader to produce a history of the last war, they had assumed the story would be limited to the war itself and, especially, to his premiership. Instead, Churchill's first volume chronicled his failed efforts to warn of the looming Nazi danger. He aimed his harshest criticism at the Tory party and its leaders, who had not only ignored but disparaged him. In Churchill's telling, he was the hero of the tale and fellow Conservatives emerged as its villains. Following the 1945 general election, public displeasure with the shortsightedness of the prewar regimes of Stanley Baldwin and Neville Chamberlain had been one of the factors generally assumed to have contributed to the Conservative defeat. Three years later, at a moment when Conservatives had reason to believe they could return to power sooner than previously expected, Churchill had focused renewed attention on the party's mistakes in the 1930s. Tory upset with this approach was by no means confined to those, such as Lord Halifax, who had themselves embraced the policy of appeasement. There was also much distress among the party's young bloods. They feared that their leader had wantonly and selfishly put their chances in a new general election at risk.

The extracts revived the charges of self-centeredness and of a fatal lack of judgment that had clung to Churchill for decades. Had not the party's belief that Churchill's prestige and popularity would be enough to assure victory been one of the factors behind the cataclysm of 1945? And now, by enhancing his personal stature at the expense of his party's, had the Conservative leader not ensured that Tories would again have to depend on the glamour of his name and reputation to carry them back to power?

The week Churchill's book, *The Gathering Storm*, was published in the US, tensions over Germany, which had been escalating since the foreign ministers' talks, came to a head. On June 24, in response to an American, British, and French decision to merge their occupation zones and to introduce a new currency that would render an independent Germany economically viable, Stalin cut off road and rail access to Berlin. The long-feared blockade had become a reality.

Churchill had often been accused of the rhetorical trick of exaggerating dangers, of always seeming somehow to claim that it was "the hour of fate and the crack of doom." The Berlin blockade gave point to his argument about the parallels between the run-up to the Second World War and the present. Speaking at a mass rally after the blockade went into effect, Churchill noted that the issues at stake were as grave as those in play at Munich ten years previously. He argued that only firmness, of a sort that had not been demonstrated early on with Hitler, would be capable of preventing another world war.

That same day, Harry Truman launched the Berlin airlift. American and British fliers undertook to provision West Berlin, thereby asserting the Western powers' right to be there. As usual, it frustrated Churchill to be out of office when great events were unfolding. While he was delighted that Truman had taken action and while he quickly moved to praise the president's response, Churchill was certain that had he been in power he would have done things differently. He thought it would have been better to respond to the blockade with counter-measures against Soviet shipping and imports that might be useful for war purposes, giving the Western powers "something practical to bargain with." Nevertheless, he was emphatic in his support for the show of resolve represented by the airlift.

How would Stalin react to the airlift? Would he dare shoot down an American or British aircraft, and what would Washington's response be if he did? These questions remained in suspense when Churchill went off to Aix-en-Provence to "bulldoze" volume two of his memoirs—which was to cover Britain's "finest hour" in 1940—for publication early the following year. There was, in Salisbury's words, "much heartburning" among Conservatives when their leader accepted an invitation to stay at La Capponcina, Lord Beaverbrook's villa at Cap d'Ail near Monte Carlo. The Canadian-born press lord, who had served as minister of aircraft production in Churchill's wartime government, had bitterly withdrawn from politics after the general election and now spent much of the year abroad. Beaverbrook remained a controversial figure in a party that still blamed him for, among many other sins, its disastrous electoral strategy in 1945. It had been "the Beaver," as he was known, who had encouraged Tories to see Churchill as their sole asset. And it had been he who had encouraged Churchill (as if he had needed any encouraging) to target the socialists with invective. Three years later, Churchill's sojourn at La Capponcina caused his party to fear that he was again in league with Beaverbrook.

Salisbury, who, with Eden, had long disdained Beaverbrook as an adventurer, denounced Churchill's stay with him as a "criminal error." To add insult to injury, when Churchill returned from his five-week working holiday for the yearly Conservative conference, he emerged from Beaverbrook's private plane.

Churchill had come home in time for the British publication of his book. It was said that when *The Gathering Storm* appeared in London shops every copy was snatched up by evening. The following day, the train from London to the seaside resort of Llandudno, Wales, was filled with delegates to the party conference carrying the book. When Churchill wound up the proceedings on October 9, he used the occasion to explicitly tie the subject matter of his memoir to the combustible crisis in Berlin in 1948. He reminded his audience that it had not only been in the 1930s that his warnings had been ignored. He recalled that in 1945 he had similarly pointed out the error of allowing the Russians to take Berlin and that he had opposed pulling back American and British troops in Germany until a postwar settlement was secured. Churchill expressed wonderment that it had taken the British and American people so long to perceive the Soviet threat after having made the same mistake before with Hitler. So, when he offered advice on how best to handle present problems, he asked his audience to consider an undeniable fact: "I have not always been wrong."

The speech did not play quite as Churchill had expected. Bursts of applause interrupted his presentation, but there was an undercurrent of disappointment among those Conservatives who wished that Churchill would spend less time focusing on a coming war and more on the 1950 election. Many members had been eager to hear a statement of Tory beliefs that might serve as a manifesto. They wanted him to comment on domestic policy. They expected him to offer specifics about what Conservatives would do to improve people's lives in an era of hard times. Instead, Churchill spoke entirely of the crisis with the Soviet Union, which, as Salisbury complained, was not at present an issue between the Conservatives and the government. Afterward, Viscount Kemsley, the newspaper proprietor, gave voice to the discontent with Churchill's leadership when he publicly called on him to share his responsibilities with younger colleagues. Kemsley was widely understood to mean Eden.

The Conservatives were not alone in criticizing their leader's speech or in viewing it in terms of party prospects. On the basis of his remarks in Llandudno, Moscow radio declared it obvious that Churchill hoped "to return to power on the crest of a new war." Three years after Stalin had incorrectly called the 1945 British general election, he predicted that Churchill would again be rejected by his people, for whom the horrors of the Second World War were still fresh.

Eisenhower entered the fray the following month with the appearance of his war memoir, *Crusade in Europe.* He had begun the book the previous February at a time when the brewing Berlin crisis made it likely that certain of his wartime decisions would be reevaluated in light of postwar realities. Dictating at a pace of some five thousand words per day, he produced a manuscript in forty-six days. The book cited a range of wartime strategic disagreements with Churchill and the British. On the sensitive subject of Berlin, Eisenhower vigorously defended his refusal to listen to Churchill about the need to take the Reich capital before their Communist allies did. Eisenhower maintained that military plans should be "devised with the single aim of speeding victory" and that the sort of political considerations expressed by Churchill ought to have had no place in his own calculations.

In fact, Eisenhower argued, it was Churchill who had made a terrible mistake with far-reaching repercussions, when he refused to listen to Eisenhower about the cross-Channel invasion. According to Eisenhower, Churchill's pessimism about Operation Overlord had resulted in the fatal error at Yalta of setting the occupation line in Germany too far to the west. At Llandudno, Churchill had hit hard at the American miscalculations which, in his view, had contributed to the current crisis. In Eisenhower's telling, it was really Churchill's fault that the Western allies had missed the critical opportunity to occupy more of Germany.

Speaking in the House of Commons on December 10, Churchill

said that he preferred not to respond to Eisenhower until he published the relevant volumes of his own war memoirs: "I did not always agree with General Eisenhower on strategic decisions, and I shall take an opportunity of expressing my views if my life—and the life of the government—are suitably prolonged." The chamber erupted in laughter at the half-bantering suggestion that Churchill might have to put off completing his memoirs if the Conservatives regained power in 1950.

He certainly hoped to have to put it off. After Christmas, as Churchill left for a short holiday in France, he sent a New Year's message to party members expressing the wish that 1949 would prove to be their "year of destiny." But a Labour triumph in the bellwether South Hammersmith by-election in February 1949 prompted renewed calls, from Kemsley and others, for Churchill to shed at least some of his power in anticipation of handing over the leadership altogether. The by-election had been deemed so important that, at the behest of campaign managers, Churchill had toured the London borough in an open car on the eve of the poll. As in 1945, the crowds who came out to welcome their beloved wartime leader did not necessarily cast their votes for his party. Despite the austerity that continued to grip Britain, Labour romped to victory in South Hammersmith. The Conservatives had been highly optimistic about their chances of winning back the seat, and the defeat occasioned what was said to be the worst crisis of faith in Churchill since he left office three years previously. There was much feeling in the party that as the general election drew near, Churchill's aloof, absentee, haphazard leadership was no longer tolerable.

On the evening of Thursday, March 3, the 1922 Committee conducted an inquest into the by-election loss that key party members were blaming outright on Churchill. Unperturbed, Churchill made a show of savoring a cigar for over an hour as his critics aired their complaints. When at last he rose to reply, he was all optimism and good cheer.

Were his colleagues downhearted? He coolly insisted that the outcome in South Hammersmith in no way contradicted what he saw as a national trend toward the Conservative Party. Were they pessimistic? He asserted that there was no reason to feel so and that Labour was sure to suffer heavy defeats later on. Did they think he was too often absent from the House? He expressed confidence that in his stead Mr. Eden gave "great satisfaction." (Like his father before him, Churchill found it impossible to enunciate the sibilant "s.") Did they wish him to spell out domestic policy details for use in electioneering? He pledged to do so before long. Did they want him to concentrate on the general election? He promised to clear his own "personal decks" and to throw his energies into home politics as soon as he returned from a crucial visit to the US.

Churchill warned the backbenchers not to be "cowed because of a bump here and there." He was so incongruously perky and positive in the face of defeat that Cuthbert Headlam, who attended the crowded session, remarked that the members were fortunate he had not given the V-sign.

At the time, Clementine Churchill seemed a good deal more concerned about her husband's troubled relations with his party than he was. It was not that she had any particular attachment to the Tory party. Clementine was a Liberal, read the Liberal press, and tended to have a low opinion of the Conservatives. On the grounds that they did not deserve Winston's help after the war, she had lamented the prospect of his using his "great prestige" to return them to power in 1945. And she certainly did not hope to see him become prime minister again in 1950. As much as any of his Tory critics, she would have liked him to give up the party leadership before the next general election. As much as any of the men who had plotted to oust him, she was eager that the succession finally occur. But she wanted him to be able to hand over on his terms, not because he had been pushed out, or because the party had grown disenchanted with him. She wanted the decision to be his alone.

So long as her husband remained in public life Clementine was ready to put her own interests aside. Though she suffered intermittently from low spirits and poor health, and though she had often felt, in her daughter Mary's words, as if she would "gladly exchange the splendors and miseries of a meteor's train for the quieter more banal happiness of being married to an ordinary man," she did what it took to help him prevail. Visiting his constituency and attending to his constituency correspondence were the least of it. In important ways, Clementine kept her ear to the ground as he simply did not.

To counter the oft-heard charge that her husband held himself apart from all but a few members of his party, she presided over numerous small, elaborately choreographed luncheons in the dining room overlooking the walled rear garden at Hyde Park Gate, where backbenchers whose names Churchill might otherwise be unlikely to remember could be made to feel that they were on intimate terms with the leader and his wife. "Clemmie" Churchill warmly greeted them at the front door and during the course of the meal "Winston darling," as she referred to him, talked confidingly of politics new and old. He even sang a song or two. "Very successful," the MP Henry "Chips" Channon noted in his diary after one such luncheon, "but I still wonder why I was asked?"

Two days after her husband faced down the 1922 Committee, Clementine wrote the first of a pair of letters to warn him of an impending problem. She often put such ideas on paper even when she and Winston were living under the same roof. He believed he had already done more than enough to pacify his critics in the party. Clementine disagreed. When the backbenchers broached the matter of his absences from Parliament, he had assured them that his trip to the US promised to be of great benefit to the party. He naturally preferred not to speak of his plan to enjoy a holiday afterward in Jamaica, where Beaverbrook had invited the Churchills to stay at Cromarty House, his residence near Montego Bay. Clementine had long disapproved of her husband's friendship with Beaverbrook, whom she thought dissolute and disreputable, but in the present instance her concern was not so much personal as political. She warned that Winston's acceptance of his hospitality would come off as cynical and that it would further alienate Conservatives at a moment of considerable doubt and discouragement among them. "I do not mind if you resign the Leadership when things are good," she said, "but I cannot bear you to be accepted murmuringly and uneasily." In the interest of averting what threatened to be construed as an insult to the party she persuaded him to send his regrets to Beaverbrook.

Churchill sailed to America in mid-March in the belief that he and Truman were to speak on consecutive days at the Massachusetts Institute of Technology. He hoped to repeat the impact of their joint appearance in Fulton, but before his ship reached New York the White House announced that the president would not be able to speak at MIT after all. British Foreign Secretary Bevin was due in Washington to sign the North Atlantic Treaty and, though it was not stated publicly, there was concern that Truman's sharing a podium with the Tory leader might cause distress at Number Ten.

Following this disappointment, Churchill had some very good news when he returned to England the following month. On April 8, as the *Queen Mary* docked eighteen hours late due to heavy gales, he received word of a surprise Tory success in London's municipal elections. The mood in the party was one of jubilation. Lord Woolton, the national chairman, crowed that the municipal elections offered a more accurate picture of broad public opinion than any parliamentary by-election. Suddenly, Churchill's calm, confident assertion that there was a national trend toward the Conservatives did not seem so far-fetched after all, and the way seemed clear for him to lead the Tories to victory. At any rate, he was the leader they had, and they were unlikely to find another in time for the general election.

In Conservative circles, the election upset temporarily silenced a good deal of the most conspicuous "anti-Winstonism." The news that Eden, overstrained by his several responsibilities, had fainted not once but twice in the course of an address to the United Nations Assembly made it difficult to argue, at least for now, that the party would perform better under the younger, supposedly more vigorous man. Overseas as well, events seemed to arrange themselves in

Churchill's favor. On May 9, the Soviets lifted the Berlin blockade. In the face of strength, Stalin had backed down from his bluff. The Soviet leader was left in what was widely viewed as a dramatically weakened position—exactly where Churchill wanted him when they met again at the conference table, which, he hoped, might be very soon.

Despite his vow to refocus on home politics, Churchill returned to the international stage in August. In Strasbourg, he participated in the first sessions of the Consultative Assembly of the Council of Europe. After he had spoken on the need to organize Western Europe against the threat of tyranny, he flew to Monte Carlo, where he planned to write, paint, and relax for a few days before he returned to the assembly to press for the inclusion of the Germans in the new European community. Churchill's exultant mood in Strasbourg—"like a schoolboy," Macmillan described him—persisted at Cap d' Ail. Beaverbrook, his host, had recently turned seventy and had proclaimed that he intended to give up neither his passions nor his prejudices nor his work. On the hot, windy afternoon of August 23, Churchill performed somersaults in the sea for the entertainment of the actress Merle Oberon. After four years, everything pointed to the political comeback that had once seemed impossible. The cup was almost at his lips.

That night in his bedroom at La Capponcina, Churchill, aged seventy-four, had a stroke.

X

THE DAGGER IS POINTED

La Capponcina, Cap d'Ail, France, August 1949

August 23, 1949, was to have been Churchill's last night in Cap d'Ail. The first sign that anything might be wrong came at about 2 a.m. He and Beaverbrook's friend, Brigadier Michael Wardell, were sitting up late playing gin rummy after dinner. Suddenly, Churchill heaved himself up from his chair. As he did, he reached out to steady himself on the edge of the table. He bent his heavy right leg a few times, as if it had fallen asleep. Complaining of cramps in his arm and leg, he clenched and unclenched his right fist. He sat again and finished the card game, though the cramps persisted.

"The dagger is pointed at me," he said before he retired for the night. "I pray it may not strike." Uncharacteristically, he asked an aide to remain with him in his room while he bathed.

On the afternoon of August 24, Churchill was scheduled to return to Strasbourg where he planned to dine with Macmillan and some others whom he had left to pursue the German issue. But when he awakened at about 7 a.m., he found that his arm and leg were still troubling him. He tried to write his name, but could not form the letters with facility. Beaverbrook summoned a local doctor. Finding that Churchill had suffered a stroke, the physician asked Moran to come without delay.

In Marrakesh the year before, when Churchill developed a fever and a hacking cough, the sudden arrival from England of his wife and his personal physician had set off what amounted almost to a death watch in the international press. The daily reports on Churchill's condition had caused much alarm to his friends and supporters at home. The news coverage, not to mention the swarm of reporters and photographers who descended on his hotel, had also greatly upset Churchill. Above all, he had worried about the repercussions to his political career of all that newspaper ink devoted to his ill health.

This time, Moran arrived at the airport in Nice lugging a set of golf clubs, and Clementine Churchill, though anxious about her husband and eager to join him, stayed away lest the sight of her trigger alarms. Moran's conclusion was the same as the local doctor's. It had been a stroke, a slight one fortunately, but a stroke nonetheless. Churchill's first reaction was to ask if he was likely to have another, a fear that would haunt him from then on. He spoke of the general election and of the possibility that he might soon be called on to lead the country. He managed a grin and suggested that he felt as if he were balanced between the premiership on one hand and death on the other. He insisted, as in the past, that he was not worried and that fate must take its course. He struck the doctor as calm, but also perhaps a bit fearful.

Moran reassured him that the episode had been limited to a small clot in a small artery. The artery had not burst and there had been no hemorrhage in the brain. Neither his speech nor his memory appeared to have been affected, and there seemed to be no paralysis. Churchill did, however, complain of a veil that seemed somehow to separate him from things. He reported an uncomfortable sensation of tightness over both shoulders. He was also concerned that his physical appearance would alert people to the fact that he had suffered a stroke, and that his bid to regain the premiership would thereby be doomed. It was not just the electorate he had to worry about; it was his party as well. According to Churchill's doctor, for now the best—and the only—thing he could do was rest.

On August 25, Moran released a statement that the Conservative leader had contracted a chill while bathing on the Riviera. At the Council of Europe, rumors predictably flew that Churchill was gravely ill. Some said he had developed pneumonia, others that he had had a stroke. On the morning of the announcement, Churchill's son-in-law Duncan Sandys, Diana's husband, informed Macmillan that he had spoken to Churchill on the telephone and that he had indeed caught a chill. According to Sandys, he was in fine form, but was unlikely to be permitted to return to the assembly for some days at least, if at all. The report, though comforting, did not altogether dispel the rumors.

For the first three days after his stroke, Churchill remained in seclusion at La Capponcina, where he practiced signing his name obsessively. Again and again, he showed his efforts to a secretary and asked whether the signature seemed normal.

In the days that followed, he tested himself in other ways. He had to find out if he could show himself in public. He had to know if his secret could be kept. One lunchtime outing, at the Hôtel de Paris in Monte Carlo, went well. As Churchill entered, the other diners rose respectfully, and no one seemed to notice that he had some trouble walking to his chair. Dinner the next evening was less successful. He was in a touchy mood, and when a woman at a neighboring table asked him to autograph her menu he could not hide his anger.

At times, Churchill seemed to think that he would be able to put the stroke behind him. "The dagger struck, but this time it was not plunged in to the hilt," he told Wardell. "At least, I think not." At other times, he behaved as if he knew that he had had his "notice to quit the world." To an aide, he spoke sadly of his four "wasted" years out of office, fighting his way back to power when there had been so much else that needed to be accomplished before it was too late. Would he be able to accomplish any of it now? For the moment, he was not even sure of his ability to walk properly when he got off a plane in England.

Churchill was eager to get home, but he worried that the inevitable press contingent at the airport would spot that something was amiss. He was so nervous about faltering in public that when he flew back on September 1, he furiously gestured to the photographers waiting at RAF Biggin Hill to leave him alone.

For the entire month, he continued to assess his ability to carry on. The question for Churchill was never whether he should continue in politics, but whether he could. The answer depended in part on his ability to conceal what had happened at La Capponcina. Significantly, both key witnesses were prepared to keep his secret. Unlike Smuts and King, who had previously advised Churchill to retire for his own sake, Beaverbrook and Moran were convinced that retirement would almost certainly be the death of him.

Nine days after Churchill returned to England, he made a first sustained public appearance that involved a good deal of walking. Lest he stumble, Clementine Churchill hovered protectively at his side when he attended the Lime Tree Stakes to cheer on his gray three-year-old, Colonist II. The Churchills arrived three-quarters of an hour early to allow the former prime minister to show himself on the clubhouse balcony to an enthusiastic crowd below. After his horse outran four rivals, Churchill set off to visit the paddock. On his way there, he was pleased to hear one racing fan call out, "I hope you can win the election just as easily!" Days like this bolstered his confidence immeasurably.

He also had dark, despairing days. A meeting in London with Sir Stafford Cripps, the chancellor of the exchequer, was one of the worst. It had fallen to Cripps to inform Churchill of the impending devaluation of the pound. The government's decision, and the spike in prices it threatened to produce, suggested that Labour would soon have little choice but to submit their case to the nation. Churchill for his part could count nine separate occasions on which Cripps had said he would never devalue. As recently as July, the chancellor had claimed that he would prefer to die first. Churchill's encounter with him therefore should have been a pure triumph for the Conservative leader. Instead, when the two men faced one another on September 18, it was Cripps who managed to seem cool and in control and Churchill who was palpably nervous.

Afterward, Churchill blamed his inadequate performance on his health. How could he fight an election, no less resume office and take on Stalin, if he felt like this?

At a moment when Churchill was weighing the possibility of handing over the party leadership after the general election, major news arrived from Washington. On September 23, Truman announced that an atomic explosion had occurred in the Soviet Union sometime within the past few weeks. The news that Moscow had acquired the secret of the bomb galvanized Churchill. He had warned of precisely this development when, in 1945, he addressed the House of Commons on the need to reach a postwar settlement while America alone had the bomb. Now, the Soviets had tested a bomb; soon they would be able to build their own arsenal, and the breathing space would be at an end. After he learned of the Soviet breakthrough, the possibility that Churchill, weakened by a stroke, would throw in his cards anytime soon was no longer seriously in play.

Ten days after the encounter with Cripps that had filled him with self-doubt, Churchill was by all accounts at the height of his powers when he held the House "entranced" for more than an hour. Officially he was there to attack the government on the matter of devaluation and to argue that Labour had brought the country to the brink of bankruptcy. But, as most listeners understood, his real purpose was to militate for an early election, a chance to get his hands back on the levers of power sooner rather than later.

And for all his talk of the financial crisis, his abiding preoccupation was clear: "Over all there looms and broods the atomic bomb."

Churchill's assault on the government was greeted with catcalls from the Labour benches. Speaking in a high-pitched voice with a Welsh lilt, the podgy, red-nosed minister of health, Aneurin Bevan, who had harassed Churchill brutally during the war, called on him to retire at once before he damaged his reputation further. Bevan predicted that should the Conservatives get back in they would promptly fling their leader aside "like a soiled glove." Nevertheless, there appeared to be broad consensus in the House that, with Churchill's speech, a watershed had been reached and an election would probably take place in October or November.

The Tories gathered in London at their annual conference in the expectation that Attlee was likely at any moment to advise the King to dissolve Parliament. Instead, the prime minister surprised them by declaring that there would be no election until the following year.

On November 30, Churchill turned seventy-five. In contrast to his gust of anger at the airport two months previously, he gamely posed for photographers at his front door in Hyde Park Gate.

"I hope, sir, I will shoot your picture on your hundredth birthday," said one member of the press.

"I don't see why not, young man," Churchill replied. "You look reasonably fit and healthy."

Did Churchill at his great age have any fear of death? "I am prepared to meet my Maker," he answered. "Whether my Maker is prepared for the great ordeal of meeting me is quite another matter."

Publicly, Churchill still claimed to have no immediate plan to retire. In private, he told a different story. He informed his friend Brendan Bracken, who passed the information on to Beaverbrook, that in the event the Conservatives lost the 1950 general election he would indeed step aside as party leader.

On New Year's Eve, the Churchills traveled to Madeira, Portugal, where Winston intended to work on his Second World War memoirs until about mid-January. On the 10th, word reached him that Attlee had given in at last. Polling day was set for February 23, 1950. This was the chance that, given the mathematical realities of Churchill's age and the size of the Labour victory in 1945, it had once seemed inconceivable he would ever have again. "I heard there was going to be a general election," said Churchill, in a mellow mood, on his early return to London, "so I thought I had better come back in case I was wanted."

The election campaign would begin officially on February 3 when the King dissolved Parliament. In preparation, Churchill plunged into a crowded schedule of "toil and moil," as he reported to Clementine, who had remained behind in Portugal. Was it all not too much for a man who had suffered a stroke five months previously? He struggled to reassure her that his "immense programme" was not more than he was capable of. He was acutely aware of the risks, but as he told Violet Bonham Carter, who visited him at Chartwell on January 16, "When great events are moving, one must play a part in them." He would probably never have another such opportunity again and he was determined to seize it with all his strength.

On one occasion, he and Rab Butler, whose expertise was in domestic affairs, worked all evening preparing the party manifesto. Finally, Churchill looked up and said, "It must be nearly eleven, we must think of moving." In fact, it was 4 a.m. Another day, Churchill and his team worked for nine consecutive hours at the massive oak dining table at Hyde Park Gate.

On the morning of January 24, Churchill was working at Chartwell when suddenly everything went misty. He could still read, but with difficulty. Frightened that this could be the prelude to another stroke, he summoned his doctor. Moran diagnosed an arterial spasm, a disturbance of the cerebral circulation brought on, almost certainly, by overwork. He suggested that Churchill could expect to have other arterial spasms if he persisted in exhausting himself.

The official launch of the campaign was still eleven days away. The travel, the speeches, the cut-and-thrust of a hotly contested election had not even begun. Yet already his body was emitting unambiguous danger signals. Churchill's instinctive response was one of defiance. As he once said, "I am not a submissive man." He did not easily give in to other people, to circumstances—or even to his own body. No sooner had he finished conferring with his doctor than he plunged back into the maelstrom of activity. Later, he ignored Moran's suggestion that, as much as possible, he substitute radio broadcasts for public appearances.

Churchill claimed to be invigorated by the crowds—admirers and hecklers alike. But the fact was that, as the electioneering began, he did appear to be markedly off his game. The problem at this point was not so much physical as it was psychological. More than ever, the possibility that he might have another, perhaps more serious stroke shadowed his every move.

Overall, the Conservative campaign opened weakly. National polls that suggested an upswing in support for Labour candidates were causing many frowns at Tory headquarters. A mood of defeatism afflicted the Conservative rank and file. In his remarks to voters throughout the nation, the Conservative candidate for Woodford dutifully concentrated on bread-and-butter issues. These were the matters Churchill had been assured repeatedly that people cared about most, yet his oratory never really caught fire.

Across the Atlantic, Truman had announced that the US was developing a hydrogen bomb more powerful than the weapon that had destroyed Hiroshima and Nagasaki. The president's message was meant to reassure Americans and their allies that their side still enjoyed superior strength. Britons, however, found the news, which came on top of some much-publicized Soviet chest-thumping about their new weaponry, anything but calming. For a broad swath of voters, the prospect of another world war even more destructive than the last overrode all other issues. The outcome of the 1945 general election had fatally interrupted Churchill's plan for a diplomatic showdown with Stalin. Five years later, he sensed that the electorate, horrified by the arms race, might be ready to send him back to the mat.

Churchill was hardly alone in perceiving that a shift in public opinion about national priorities had begun. He was, however, the first politician to tap into it. He sculpted a speech, to be delivered in Edinburgh, calling for a new Big Three conference, and the Tory hierarchy found themselves counting on his words to reenergize the party's campaign. The day before Churchill's address, Anthony Eden, speaking in London, signaled the world to pay particular attention to what Churchill was about to say. Privately, Eden had long maintained that Churchill was far from indispensable and that he, Eden, was ready and amply equipped to replace him. He regarded Churchill as delusional for believing that he alone could head off a war with the Soviet Union. He believed that Churchill suffered from what amounted almost to an infatuation with Stalin. He was convinced that Churchill had actually done a poor job at Potsdam. And, he also thought foreign ministers' talks preferable by far to the top-level contacts Churchill favored. At the moment, however, Churchill's call for a rematch with Stalin did appear to be the Conservative Party's best chance to return to power. Eden therefore played his part by lauding Churchill as the only world leader capable of ending the Cold War and by suggesting that a Conservative victory would lead to a bold new initiative to avert an atomic war.

On February 14, Churchill began by delivering some of his usual blood-and-thunder anti-socialism to an overflowing crowd at Edinburgh's Usher Hall. Thousands who could not gain admittance stood outside in the rain listening to his remarks on loudspeakers. During the speech, Churchill's attack mode dissolved and his voice became grave. He conceded that he was privy neither to the government's secret information nor to US intentions.

"Still," he went on, "I cannot help coming back to this idea of another talk with Soviet Russia on the highest level. It is not easy to see how things could be worsened by a parley at the summit, if such a thing were possible."

The effect of his words on the general election campaign was electrifying. To the government's chagrin, his call for direct talks with Stalin—a "summit" conference, as top-level discussions came to be called—proved immensely popular, but by no means with everybody. Foreign Secretary Bevin dismissed his proposal as an electioneering stunt. Sir Stafford Cripps scoffed at the idea that Churchill-Stalin talks would bring salvation to the world, and another prominent Labour politician assailed Churchill's brand of "soap-box diplomacy."

And that was just what was being said openly. Whispers emanated from the Labour camp that Churchill would have the nation at war within eighteen months of his return to Number Ten, that he was motivated not by any true love of country but rather by a lust for power and a craving for retribution, and that he was simply too old and ill to form a new government.

Finally, some forty-eight hours after Churchill's Edinburgh speech, rumors swirled that he was dead.

On February 16, Churchill commented humorously from Chartwell, "I am informed from many quarters that a rumor has been put about that I died this morning. This is quite untrue." In a radio address the next day—his last of the campaign—he sought to turn the issue of his seventy-five years to his advantage. "Of course, as I am reminded, I am an old man," Churchill said. "All the day dreams of my youth have been accomplished. I have no personal advantage to gain by undertaking once more the hard and grim duty of leading Britain and her Empire through and out of a new and formidable crisis. But while God gives me the strength and the people show me their good will, it is my duty to try and I will."

On polling day, Churchill visited his constituency before returning to Hyde Park Gate to await the results in the company of Salisbury and other prominent Conservatives. In the last lap of the campaign, Salisbury, though he had previously opposed further talks with Stalin on the grounds that they would be akin to the negotiations with Hitler before the war, had spoken out to support Churchill's "summit" proposal. If the Conservatives managed to regain power, there was acknowledgment in the upper reaches of the party that it would be largely thanks to Churchill.

There was also a sense, both on Churchill's side and on that of the men who had previously worked to oust him, that the 1950 general election was his last chance. Even Churchill seemed to accept that he could hardly afford to wait another five years to begin a new government. By the time another full election cycle passed he would be eighty, so it was now or never. If the Conservatives were unsuccessful tonight, he would have little choice but to finally hand over.

As the evening began, it appeared that Churchill was already finished. Early reports showed strong figures for Labour. Then matters began to improve as Conservatives scored heavy gains. The vigil stretched into morning, when it seemed as if Labour might have lost its majority. But by noon on the 24th, it was clear that the socialists had managed to hold on to power after all. This time, however, their majority over all parties combined was only six—hardly enough to maintain anything like a stable government.

Among Conservatives, there was frustration at having come so close to winning, but there was also elation that another Labour landslide had been averted. Five years previously, the wisdom had been that the socialists would be invincible for at least a decade. That apocalyptic assessment had been disproved. Lord Woolton, who had done much to reorganize and revitalize the party in the wake of 1945, called the 1950 general election "a moral victory for the Conservatives if there ever was one."

Because of the precarious Labour majority, the contest was immediately and universally regarded as, in Macmillan's words, "the prelude to another round." Less certain from the Tory point of view was when that next round ought optimally to take place.

Churchill believed that another general election in the next few months was inevitable. In light of that estimate, he was quick to reconsider his previous intention to shut up shop in the event of a Labour victory. Five years was one thing, a few months another. To the embarrassment of Conservative colleagues, it soon became apparent that, whatever he may have suggested beforehand, he had every intention of continuing as party leader.

Cheered on by Macmillan, who urged him to press for a government defeat on an issue of major policy within three weeks of the opening of the new Parliament, Churchill aimed to force an election as soon as possible. He calculated that it could take as many as three, but probably no more than six months. In the first month alone, he would lead four determined efforts to overthrow the government.

In Conservative quarters, Eden questioned the wisdom of Churchill's determination to go against Labour so hard and so fast. Part of this, no doubt, was Eden's sincere distaste for pugilistic politics. He had long striven to distance himself from Churchill the "political warrior." In the present instance, however, Eden targeted not just Churchill's tactics, but also his motives. A few weeks after he had hailed Churchill as the only world leader capable of ending the Cold War, the leader in waiting could not comfortably argue that Churchill ought to hand over at once when a new election and conceivably a meeting with Stalin were only months away. Eden made the case that Churchill, mindful of his age and declining powers, was aware that unless he could get the government out quickly, time would prevent him from ever becoming prime minister again.

There was truth to the charge that Churchill was prepared to allow his personal timetable to dictate political decisions. When Churchill spoke of the need to act before it was too late, it was sometimes hard to know whether he was referring to the world situation or his own. For an "old man in a hurry," the two can seem perilously interchangeable.

Yet, in arguing for a more cautious approach, Eden was not exactly disinterested either. Just now, it served him to proceed slowly as much as it served Churchill to move quickly. The longer the Labour Party managed to hang on to power, the less chance there was that Churchill would still be in place to carry the flag at the next election.

XI

ANOTHER GLASS OF YOUR EXCELLENT CHAMPAGNE

Venice, 1951

On August 23, 1951, a train carrying Winston and Clementine Churchill, along with their fifty-five suitcases and trunks, was approaching Venice, where guests had already begun to arrive for the masked ball at the Palazzo Labia that promised to be the most extravagant party in all of Europe since before the Second World War. Lady Diana Cooper was going as Cleopatra; Deborah, Duchess of Devonshire, as Georgiana, the wife of the fifth Duke; and the American socialite Barbara Hutton as Mozart. The evening's flamboyant Mexican host, Charles de Beistegui, intended to preside over "the party of the century" in the garb of a Venetian procurer, with flowing red robes, a curly wig, and platform soles that increased his height by sixteen inches.

Many of the guests stayed at the Excelsior Palace Hotel on the Lido, which was also swarming with visitors who had neither received one of the 1,500 invitations nor purchased one on the black market but hoped that one might yet be had. The aspirants need not have bothered to come all the way to Venice. For weeks, bona fide invitees had been begging to be allowed to bring guests of their own. Beistegui replied indignantly that if he let in everyone who wished to be admitted, his eighty-nine-room palazzo, which housed Tiepolo's frescoes of Antony and Cleopatra, would sink into the Grand Canal.

There was to be an official welcoming ceremony for the Churchills at the train station, after which they were due to proceed by motor launch to the Excelsior. Shortly before they arrived, Clementine Churchill left the compartment to prepare. In her absence, Churchill, who hoped to do a bit of painting in the city of the doges, leaned out the carriage window for a better view.

He had been discussing art materials with his bodyguard when suddenly the latter rushed forward, grabbed his right shoulder, and pulled him back in. A fraction of a second more and Churchill's head would have hit a concrete post that held up some electrical wires, only a few inches from the side of the train. When Churchill grasped what had just happened, he said with a smile, "Anthony Eden nearly got a new job then, didn't he?"

Eighteen months after the general election, British political life remained deeply unsettled. In spite of Churchill's confident prediction in February 1950 that a new election was imminent, his efforts to bring down the government had so far come to naught. Labour retained its gossamer hold on power. Churchill compared the prolonged physical and mental strain of waiting for an election that might be called at any time to "walking under a tree with a jaguar waiting to pounce." George VI, meanwhile, had begun to question how much longer Churchill could possibly go on in politics. Though he had been opposed to Churchill in 1940, he had become devoted to him during the course of the war and had been stunned and dismayed when the people turned him out in 1945. Nevertheless, several times in recent months, King George had conveyed his anxieties about Churchill to Eden. On one such occasion, to the fascination of onlookers at a reception at Buckingham Palace in February 1951, the King conferred for a good twenty minutes with the Conservative heir apparent. The scene prompted suspicions that a new alliance might be in the making, one that could mean trouble for Churchill.

Seeking to counter sentiment that he was already too old to form a government, Churchill led the opposition in turbulent all-night sessions in the House of Commons in June, including one that lasted nearly twenty-two hours. To his party's delight, he voted in every division, gave marvelous speeches, and spiced his remarks with jokes and raillery. Afterward, rather than go home, he would make a show of devouring a huge breakfast that began with eggs, bacon, sausages, and coffee, followed by a large whisky and a cigar.

That, at least, was the robust image Churchill hoped to project. The *Daily Telegraph*'s Malcolm Muggeridge managed to extract from Eden the information that Churchill's health was not as good as the public seemed to believe. Eden was right, of course. Few people outside Churchill's intimate circle knew about his wartime heart attack. And few knew that he had since had a stroke, let alone that he lived in perpetual anxiety that another stroke might yet keep him from returning to power.

As it happened, Churchill was arriving in Venice two years to the day since he performed somersaults in the sea hours before the first stroke. On August 21, prior to leaving the rainy French Alps en route to Italy, he had written to ask Sir Russell Brain whether he might safely swim at the Lido. Churchill promised not to go in unless the water was well over seventy degrees, and he made it clear that he had no intention of plunging in, but rather of acclimatizing his body slowly over a period of two minutes or more. Would Brain please consult with Moran and telegraph his opinion to the Excelsior?

An army of press photographers had descended on Venice, both for the masked ball and for the Venice International Film Festival. There were many actors to take pictures of, including Vivien Leigh, Orson Welles, and Errol Flynn. But no sooner did Churchill appear in town than a picture of him on the Lido became the most ardently sought-after image. Every day, boxes filled with ornate period costumes for the masquerade were unloaded from motor launches at the Excelsior and other hotels. But the one costume every cameraman wanted a shot of was Churchill's "flaming-red" bathing trunks. A picture of the Conservative leader shirtless, with a towel draped round his opulent waist, would also do. On his strict orders, two English bodyguards and half a dozen Italian helpers kept the press away from his canvas cabana and from the strip of beach where he and Clementine swam together every day, as they had done when they came to Venice on their honeymoon in 1908.

Churchill made a point of skipping the masked ball that everyone else seemed so frantic to attend. Despite this, he was much on view during his fortnight in Venice: painting a landscape on Torcello, seated beside his wife at a Dior fashion show, dictating portions of the fifth and penultimate volume of his Second World War memoirs to a secretary as they flew along in a motor launch.

Duff Cooper, who had retired as ambassador to France and now worked for the British Film Producers' Association, saw a good deal of Churchill privately that August. Cooper had last encountered Churchill in May when they had lunch together in London, and he was struck by his deterioration since then. "I found him much older and very deaf," Cooper noted in his diary. "He depends on champagne and is morose and gloomy until he has had a good deal."

It often fell to Clementine Churchill, who was herself in delicate health after gynecological surgery that had kept her in the hospital for three weeks in May, to "run interference" for her husband, warning friends like the Coopers that he was in ill humor. When the Churchills dined with Duff and Diana at the Gritti Palace in Venice, Winston was silent and detached at first. Not even Diana, whom he adored and who adored him, was capable of bringing him out of his gloom. In years past, he would have been happy just to sit beside her and talk; he would innocently beg her to give him her "paw" or a little kiss. But there was none of that tonight. Instead of smiling with pleasure as he tended to do in Diana's company, he did little but scowl. The Coopers' son, John Julius, watched his mother grow desperate as she tried everything, but to no avail. Each painful attempt at conversation was met with a grunt. Churchill's black mood swept over the table, reducing everyone to an awkward silence, until finally he addressed his hostess.

"I shall be much better when I have had another glass of your excellent champagne."

Churchill drank. And in the course of about five minutes he was indeed transformed. Suddenly he was himself again, laughing, joking, and telling stories. He sang some music hall songs from his repertoire, and half in earnest, half in jest, he told Duff that he feared Clemmie would reproach him on the way back to the Excelsior for "lacking in dignity." The evening ended most happily. Still, there could be no avoiding the conclusion that in important ways Churchill simply was not the man he had been in 1945, 1950, or even the first half of 1951.

Soon after the Churchills returned to England on September 12, Attlee announced that he had asked the King to dissolve Parliament and that a new election was to take place on October 25. Immediately, Churchill summoned Eden and other members of the Tory hierarchy to launch the campaign. Polls suggested that the Conservatives were likely to secure a working majority of thirty to forty seats (Beaverbrook went so far as to put the number at fifty), so there was huge optimism amongst them. Understandably, there was also considerable anxiety about Churchill's fitness. As he had done many times in the past, he sought to assuage his colleagues' worries by indicating that he planned to stay in power for a limited time. He put out the word in the party that in the likely event of a Conservative victory he intended to lay down the burden within a year to eighteen months.

Nevertheless, the awkward questions kept coming. Citing Churchill's demeanor when they met in Paris on September 11, NATO Supreme Commander Dwight Eisenhower asked Macmillan point-blank whether the Tory leader was "physically able" to take up the work of government. Eisenhower suggested that it might be best for Churchill to announce that after setting up things properly in the new government he would transfer the reins "to younger hands."

Closer to home, both Churchill's wife and doctor had grave concerns of their own. Moran confided to his diary that Churchill had lost ground in recent months and was probably no longer equal to the tasks that faced him as Britain's leader. Clementine Churchill worried that were her husband to become prime minister in his late seventies, it might no longer be possible to operate "at his best." Loving him as she did, she could not bear to see that happen.

Churchill's first campaign appearance, in Liverpool on October 2, left him feeling depleted and quite glad that he had nothing on his schedule the next day. When Moran visited him afterward at Hyde Park Gate, Churchill confessed that he was no longer as sure as he had once been of his ability to see things through. By degrees, his self-confidence began to be undermined in other critical areas as well: notably the assumption, in play since 1945, that he was his party's greatest asset.

Nineteen months previously, Churchill had nearly managed to beat the socialists by appealing to voters' desire for peace. This time round, Labour hijacked the issue and used it against him. Preferring to sidestep domestic and economic matters at a moment of broad public dissatisfaction with government policy, the socialists focused on the claim that Churchill's return to power would plunge Britain into another war. The whispering campaign of 1950 burst into the open with slogans like "A third Labour Government or a third world war" and "Vote Tory and reach for a rifle, Vote Labour and reach old age." The *Daily Mirror*, which supported Labour, asked readers, "Whose finger do you want on the trigger, Attlee's or Churchill's?" And Labour Defense Minister Emanuel Shinwell sniped that it would not be long before Britain's wartime leader would seek again to show off his talents.

By harping on the warmonger theme, Labour managed to dominate and drive the election campaign as Churchill had done in 1950. They put him on the defensive, and there he miserably remained until polling day.

At first, Churchill struggled to answer the charges dispassionately. Since Fulton, many at home and abroad had concentrated on his call for strength and ignored his equally important plea for negotiations. In a radio broadcast on October 8, he sought to correct that mistake. "I do not hold that we should rearm in order to fight," he explained. "I hold that we should rearm in order to parley. I hope and believe that there may be a parley." He reminded listeners of his proposal, at Edinburgh in 1950, of a top-level meeting with Soviet Russia. He recalled that the socialists had dismissed the idea as an electioneering stunt, and he suggested that had such a meeting taken place, the Korean War might have been avoided.

On October 17, at the close of three days of electioneering in the north of England and Scotland, Churchill renewed his appeal for a meeting with Stalin. Labour shot back that far from intending to press for a peace settlement, Churchill really planned to use a summit meeting to provoke another war. The *Daily Mirror* ran a story claiming that Churchill had plans to issue an ultimatum to Stalin.

By this time, Churchill's blood had begun to boil. When Stalin called him a warmonger, Churchill had managed to laugh it off. He found that he could not be so blithe when his own countrymen repeated the accusation. As he never tired of arguing, had they listened to him in the 1930s, the Second World War might never have had to be fought in the first place. Then, as now, he had been a proponent of strategic thinking as a means of preventing war. Nonetheless, the warmongering campaign seemed to have taken hold with voters.

On October 20, when a pollster visited Hyde Park Gate to review the Conservatives' chances of success, he found Churchill drinking a whisky and soda in bed at half past ten in the morning. This being Churchill, there was certainly nothing unusual about that. (He had long claimed that he drank in the morning strictly to "moisten" his throat.) What was out of the ordinary was the urgent question he put to his visitor, after the latter cautiously suggested that an overall Tory majority of about forty seemed probable.

"Do you think I am a handicap to the Conservative Party?" Churchill asked in a low, sober tone of voice.

The pollster was silent at first, but Churchill reassured him that he need not be afraid to give an opinion.

"Well," said the pollster, "I do not think that you are the asset to them that you once were."

Forty-eight hours before polling day, Churchill gave vent to his anguish over the "cruel and ungrateful accusation" that, he was convinced, had caused his stock to plummet. Speaking to an audience of six thousand in Plymouth, Churchill made it clear that he was so frustrated and upset because he believed himself to be the very opposite of a warmonger. At Fulton and on subsequent occasions, Churchill had claimed that at his age all of his personal hopes and dreams had long been fulfilled. Today, however, he confessed that he had one great ambition still. The man who had led Britain to triumph in the previous world war pleaded for the opportunity to try to avert another.

"It is the last prize I seek to win."

Churchill was furious when the *Daily Mirror* responded by repeating the "Whose finger on the trigger?" theme on its front page on polling day. There was speculation at the time that the newspaper cover alone cost the Tory party a great many votes. In any case, when the first figures came in that evening, it was apparent that the generous lead the Conservatives had enjoyed in the polls had evaporated. The close race continued through the night. Not until the following day, Friday, did a Tory victory in the constituency of Basingstoke give the party a thin majority.

When Churchill learned the news, at Conservative headquarters in London, tears coursed down his face. By an act of will, he had accomplished what the world had once (quite reasonably) believed he could never do. Within weeks of his seventy-seventh birthday, he was to become prime minister again, the oldest man to lead Britain since Gladstone. To reach this point, he had defied time, ill health, the opposition of colleagues, family, and friends, the mockery of opponents, and, at last, his own festering doubts.

In the years before his first premiership, the middle-aged Churchill had had occasion to reflect on the amount of energy some uniquely gifted people are required to waste before they even have a chance to accomplish what they want to do in life. He wrote, "One may say that sixty, perhaps seventy per cent of all they have to give is expended on fights which have no other object but to get to their battlefield." When Churchill made that observation, in an essay on the British politician Joseph Chamberlain included in *Great Contemporaries* (1937), he had himself been caught up in precisely the sort of long, draining preliminary struggle to which he referred. The observation nicely described what, in the end, was to emerge as the arc of Churchill's own political career. By the time he had realized his supreme ambition of becoming prime minister, in 1940, he had spent decades fighting to reach his particular battlefield. Again, after being hurled from power in 1945, Churchill dedicated an additional six years to fighting his way back.

On October 26, 1951, the battlefield was in sight, but did he have enough strength left for the battle?

XII

WHITE WITH THE BONES OF ENGLISHMEN

London, October 1951

Hours after Churchill learned that the Conservatives were back in he crossed the red carpet at the entrance to Buckingham Palace on his way to see the King. Only a few months previously, George VI had sensed that the Churchill era in British politics might be coming to an end. This evening, the King, who was recovering from surgery for lung cancer, asked Churchill to form a new government.

Later that night, over dinner at Hyde Park Gate, Churchill made the first appointment of his peacetime administration, inviting Eden to resume his old job as foreign secretary. Eden also insisted (though Salisbury thought him "rather an ass" for doing so) on being leader of the House of Commons. It was no secret that serving in both capacities had proven a nearly impossible strain during the war. Still, at a moment when Churchill professed to be ready to hand over in a matter of months, Eden's overriding concern was to safeguard his claim to the succession by grabbing as much power as he could. He also asked Churchill to make a public gesture that would leave no doubt in people's minds that Eden was next in line. In addition to the titles of foreign secretary and the leader of the Commons, Eden wanted to be appointed deputy prime minister.

Churchill, a devout royalist, knew that the title had no constitutional validity and might be viewed as an infringement on the royal prerogative. When Attlee held that appellation in the wartime coalition government, it had referred strictly to his focus on domestic matters, freeing Churchill to concentrate on the war effort. There had been no implication that Attlee would become prime minister should anything happen to Churchill. By contrast, in 1951, Eden construed the title wholly in terms of the succession.

Eden could be a prickly and peevish character, yet in his relations with Churchill he had rarely, if ever, been as personally demanding. (He was notorious, in the words of one observer, for "bullying people who could be bullied and collapsing before those who couldn't.") It had long been a source of frustration to Salisbury that Eden, for all his plots and plans, had so often failed to convey exactly what he wanted to Churchill. Time and again, as soon as he actually had to face Churchill, Eden's courage had deserted him. Time and again, Eden had proven willing to go only so far, apparently fearful of putting everything "into one throw" and it not coming off.

What emboldened Eden now? What accounted for the new firmness, the new insistence on having exactly the jobs and titles he wanted? Churchill was visibly exhausted after an election campaign that had required him to speak nearly every day for five weeks and that had played havoc with his hearing, which was even worse than before. One witness compared his bedraggled appearance in the first few days of the new government to that of "a sea captain who had just come out of a very heavy storm and was glad to be in harbor again." Another observer went so far as to wonder in all seriousness whether Churchill might die during the new Parliament. Confronted with this diminished figure, Eden became assertive and aggressive. Confronted with this new Eden, Churchill gave him everything he asked for.

Wearing a "quilted flowered" bed-jacket, Churchill spent the better part of Saturday morning at home in bed. A sheet of writing paper with a list of names and offices printed in oversized type lay on the coverlet. A cat sprawled across Churchill's feet in lieu of a hot water bottle as he interviewed some of the men who were to be Cabinet ministers. As if he meant to exert himself as little as possible, he accompanied his remarks with small, sharp hand gestures that began emphatically at the wrist, not the arm.

The key appointment after Eden was considered to be that of the chancellor of the exchequer. The financial situation was far worse than Churchill and his party had previously thought; the country was on the verge of insolvency. Attlee, now opposition leader (though his wife had hoped he would retire), was rumored to be relieved to have handed over the mess to the Conservatives; the expectation was that they would botch the job and that Britons would soon turn back to Labour. Churchill surprised a good many people by passing over the obvious candidate for chancellor, Oliver Lyttelton, in favor of Butler. Butler himself was astonished to be offered the post, Lyttelton being his superior both in age and knowledge of finance, as well as a favorite of a number of the Tory leaders. Butler, however, had a well-established profile as a key Conservative policy-maker and as a spokesman for the party's impatient young bloods. Since 1945, he had also received a good deal of popular attention as Eden's main rival to succeed Churchill.

A preliminary list of Churchill's appointments was sent to Buckingham Palace. Immediately, both George VI and his private secretary objected to one of the notations beside Eden's name. The King was free to choose whoever he wished to succeed as prime minister. He could consult anyone he liked about the matter—or no one at all. He and Lascelles perceived the title of deputy prime minister, with all that it seemed to suggest in Eden's case, as an invasion of the royal prerogative. Lascelles went at once to Hyde Park Gate to convey the King's insistence that Eden be denied the appointment.

At lunchtime, high words were exchanged in a telephone conversation between Eden on one end and Lascelles and Secretary to the Cabinet Norman Brook on the other. Brook seconded Lascelles's view that Eden's appointment infringed on the royal prerogative. Eden, speaking from his house in Mayfair (which had once been home to Beau Brummell), angrily countered that he had been promised the title and that he saw no reason why he should not have it. Despite the vehemence of Eden's protests, Lascelles, who had had his troubles with Eden in the past, remained firm.

The deadlock persisted until Churchill intervened. As on the previous evening, the prime minister was courtesy itself to his heir apparent. He got on the phone to reassure Eden that he would have his title after all. Churchill kept his word, for that afternoon George VI approved the original list at a Privy Council meeting, and the news of Eden's appointment was duly announced.

Churchill, in the meantime, transferred his base of operations—to his bed at Chartwell. Harold Macmillan, who initially feared that he had been passed over altogether, found himself summoned to Kent on Sunday morning. Lady Dorothy, Macmillan's wife, drove him there. She strolled in the garden with Clementine Churchill (who tended to be cool to Harold, whom she did not trust) while he went in to his meeting.

Not long after he had gone upstairs to see Churchill, Macmillan, whose walk was a slow, stiff shuffle, emerged in a fit of temper. "If he wants to kill me politically, let him do it, but not this way!"

Macmillan had originally hoped to be appointed minister of defense, a position that Churchill had wasted no time claiming for himself. Instead, after all that Macmillan had done to support Churchill when few others had been willing to back him, Churchill—"in a most pleasant and rather tearful mood"—had repaid the favor by inviting Macmillan to become minister of housing. Though Macmillan, by his own account, knew nothing about the housing problem, Churchill asked him to take responsibility for fulfilling the Conservative Party's campaign pledge to build 300,000 houses a year. Few people, including Macmillan himself, believed that target capable of fulfillment. Assured that the job was a gamble and would "either make or mar" his political career, Macmillan asked to have a bit of time before he gave his answer.

In the garden at Chartwell, Macmillan discussed the offer with his wife. It was well known in the world of Westminster that Lady

Dorothy, stout and frumpy with a nest of brittle gray hair, had long been unfaithful to her husband. Her lover, Robert Boothby, was one of the adventurers in Churchill's orbit whom the Eden faction had been desperate to bar from their anti-appeasement meetings in the 1930s. The affair, which neither she nor he took any particular trouble to conceal, was still going on. Through the years, Macmillan had poured out his anguish to his sister-in-law, Moucher Devonshire (now the Dowager Duchess), to whom he confided his belief that the Macmillans' daughter Sarah had actually been fathered by Boothby. Still, for all the hurt Dorothy had caused her husband, she was a fervent supporter of his career. She adored politics, advised him sagely, and labored tirelessly and effectively on his behalf. She was a far better campaigner than Macmillan, who tended to be stiff and overly intellectual with voters.

For better or worse, Macmillan was also inclined to be extremely cautious. Dorothy, it need hardly be said, was more of a risk-taker. During the war she had immediately perceived what Churchill's offer to send Macmillan to North Africa could do for her spouse's stalled career. Now once again, she grasped that, though certain of his colleagues might be inclined to perceive the position of minister of housing as a backwater, it could in fact prove to be a tremendous opportunity. At a time when England suffered from an acute housing shortage, Macmillan's efforts, if successful, could make him highly popular with the electorate. Additionally, the job would afford him the crucial domestic experience that, in contrast to Butler, both he and Eden lacked.

On October 30, a remarkable scene unfolded in the columned Cabinet Room at Number Ten, which Churchill had once lamented he would never sit in again. Enthroned with his back to the fireplace, he met with his new Cabinet for the first time. Despite his image over the past four days of a man in decline, Churchill had moved swiftly and shrewdly to set the various aspirants against one another. The cozy picture of the teary-eyed old fellow in a bed-jacket camouflaged the characteristic ruthlessness with which he had acted to protect himself against those who hoped to replace him.

Churchill had allowed, even encouraged Eden to poison his relations with Buckingham Palace by insisting on a title that gave offense to the King. He had further checked the heir apparent by placing the latter's most feared rival in a position of power that most observers had expected to go to another man. And he had checked Butler by placing the driven, determined Macmillan in a job that would require him to fight to the knife for a disproportionate share of government funding when it was Butler's mission to reestablish financial stability.

Churchill had named Salisbury, who had been hoping to oust him since day one of the postwar Parliament in 1945, to the position of Lord Privy Seal, but by the gambit of these other appointments he had left his old antagonist little room to maneuver. (The Conservative victory also meant that Salisbury again became leader of the House of Lords.)

At the beginning of Churchill's wartime premiership, there had been many comments about his renewed vitality and sharply improved physical appearance. Now, once more, power seemed to reenergize him. In sharp contrast to his air of exhaustion in recent days, he was "in great form" today, merrily quoting Machiavelli, urging his colleagues to leave things to him, and declaring his intention to travel to the US and Canada as soon as Parliament was "up."

The Americans had been anticipating Churchill's arrival for some time. Long before the Conservative Party returned to power, the Truman administration had determined that in the event of a Tory victory Churchill was "almost certain" to try to arrange the meeting with the Soviets he had spoken of so often. There had been much internal debate on the American side about how the US ought to respond. The various reasons for and against a parley at the summit had been weighed extensively, and by the time Churchill was back at Number Ten, Washington had taken a firm decision. Churchill had regained office just as the Americans were about to enter a presidential election year. Whether or not Truman decided to seek another term it would be politically impossible to agree to talks that could subject the president and his party to a serious attack by the McCarthyites. But there was also international opinion to consider. The administration foresaw that should Churchill suddenly make a speech suggesting that he and Truman go to see Stalin, it would be difficult in view of Churchill's personal prestige for Truman to decline. Washington therefore was keen to make Churchill aware of its opposition without delay. It fell to the US ambassador in London, Walter Gifford, to warn Churchill against any "hasty public" move toward a top-level meeting before the new prime minister had had an opportunity to review the pros and, more importantly, the cons with his American allies.

Under the circumstances, the best Churchill could hope for was that the American stance might alter after the presidential election in November 1952. He was going to have to wait a year or more before there was even a chance of returning to the table with Stalin, hardly an appealing prospect to a man about to turn seventy-seven. Meanwhile, in the hope of convincing Truman of the desirability of talks with Stalin at some later date, and of reestablishing the kind of intimate bond he had enjoyed with Roosevelt during the war, Churchill was eager to get to the US prior to the onset of election fever.

Before he invited himself to Washington, he was careful to do something else first. Churchill dashed off a telegram to Stalin. By reaching out to Stalin before he contacted Truman, Churchill ensured that should the Americans object it would already be too late.

"Now that I am again in charge of His Majesty's Government," Churchill wrote to his Soviet counterpart on November 5, 1951, "let me reply to your farewell telegram from Potsdam of August 1945. 'Greetings. Winston Churchill.'" But would it really be possible, as Churchill seemed to believe, to ignore the years in between and pick up where they had left off? At Potsdam, Churchill had acted on the premise that Stalin, though far from the man Roosevelt had imagined him to be, was capable of discerning what was in his own interest and that he would negotiate based on that understanding. Was Stalin any longer able to make such judgments in 1951? He was sicker and if possible more paranoid and frenzied than when he and Churchill last met. He suffered from hypertension, arteriosclerosis, and a host of other afflictions. He had had several minor strokes, his memory often faltered, and he was liable to lose his balance or pass out at any time. When he took a solitary walk in the country, the paths were specially equipped with telephones in green metal containers so he could call for assistance if he fell ill. Stalin lamented that "cursed old age" had caught up with him and he claimed to accept that he was "finished." Yet he gave no sign of being ready to hand over. Stalin seemed as fearful of the loss of power as of death itself.

Unlike Churchill, who clutched his heir apparent with hoops of steel lest a more potent rival for the Conservative leadership materialize, Stalin was constantly anointing some new chosen one. Unlike Eden, who seemed always to be jockeying to retain the number two spot, Stalin's minions preferred to avoid it. In their case, Stalin's "persecution mania" being what it was, the designation of heir apparent was as likely to lead to a death sentence as to the succession. Molotov, who had been Eden's opposite number in 1945, had managed to survive, but he had been ousted as foreign minister and he had acquiesced when his adored wife was imprisoned on phony charges of treason. Molotov later speculated that he would probably have been killed had Stalin lived a little longer.

Before a week had passed, Churchill reported to his Cabinet that he had heard back from both Truman and Stalin. Truman invited him to confer in Washington beginning on January 3, 1952. The Old Bolshevik's reply was less encouraging. "Thank you for greetings" was all that it said. Churchill laughed that at least he and Stalin, who had had some pretty harsh things to say about him since Fulton, were again on speaking terms.

In the opening weeks of the premiership, Eden too was looking ahead a year to eighteen months, but he was still acting on the assumption (of which Churchill made no effort to disabuse him) that the handover would have taken place by then. Churchill had reneged on previous pledges to retire, but this time a good many people besides Eden wanted to believe that, having wiped out the disgrace of his 1945 defeat, Churchill sincerely intended to let go.

Eden cut a glamorous and popular figure as head of the British delegation at the tumultuous Sixth Session of the General Assembly of the United Nations in Paris that November. When he first entered the assembly, foreign ministers and diplomats from many nations rushed up from all sides to pat him on the shoulder and welcome him back to the fray. He quickly earned lavish praise in the international press for his efforts to improve the tone of the proceedings after bitter charges flew between his Soviet and American counterparts, Andrei Vyshinsky and Dean Acheson. When Britain's suave new foreign secretary calmly but firmly recommended a truce to all the shouting, he seemed the very soul of moderation and good sense.

That, at least, was his public persona. In private, Eden was anything but calm. On the eve of the speech that was to mark his return to the international stage, he collapsed in his bedroom with stomach spasms. Initially, he feared the episode might be a recurrence of the duodenal ulcer that had sidelined him during the war and had required Churchill to take over for him. Eden's aides summoned a doctor in the middle of the night. While the physician treated him with injections, Churchill and Acheson were notified of his ordeal, the latter in strict confidence. From then on, Churchill was far from the only one in the new government whose health was not as good as the public was led to believe.

In the end, on November 12, Eden managed to give his speech. It was warmly received, and most observers had no inkling that he had nearly had to cancel his appearance. Still, for those few who knew about it, the previous night's episode underscored just how unwell the fifty-four-year-old Eden really was as he launched into the punishing sixteen- and seventeen-hour days his new job entailed. When Eden had demanded the leadership in the House of Commons on top of the post of foreign secretary, Salisbury had worried that in addition to not being able to devote enough time to both jobs Eden was likely to kill himself in the bargain. Early on, in a concession to his health Eden had agreed to transfer the leadership in the Commons to Harry Crookshank. Whether that would be enough to save him was an open question.

More and more, any amount of stress or anxiety seemed capable of triggering another bout of paralyzing pain. Eden now traveled with a black tin box that contained a smorgasbord of painkillers provided by his doctor. When he required a dose of morphine, a detective in his employ was summoned to give him a shot. By the time he landed in Rome in the last week of November for the Eighth Session of the North Atlantic Treaty Organization Council, he was in such agony that at night his loud groans could be heard through the wall in the adjoining bedroom where his private secretary, Evelyn Shuckburgh, was trying to sleep.

In Rome, there was much to make Eden nervous, not least his breakfast with Eisenhower on November 27, four days after Churchill had begun to fulfill his 1948 promise to rebut Eisenhower's *Crusade in Europe* by publishing the relevant volumes of his own war memoirs. In volume five, *Closing the Ring*, which had been serialized in twenty-four prepublication installments (advance copies of which Eisenhower had seen), Churchill accused Eisenhower of inaccuracy when he portrayed him as having lacked confidence in the cross-Channel invasion. The debate was about much more than whether or not Churchill had believed that Overlord could succeed. His skepticism then was the linchpin of Eisenhower's argument that it had been Churchill's fault that the lines of occupation in Germany had been set too far to the west. But the battle for history's verdict on who had been principally responsible for some of the worst problems of postwar Europe was far from finished. Volume six of Churchill's memoirs promised to chronicle his differences with Eisenhower and Truman over Berlin and the pre-Potsdam troop withdrawal respectively, and to make the case that he had seen the Soviet danger in time but that they had refused to listen.

Eisenhower's reaction to *Closing the Ring* was fury. In his current role as NATO's military chief, he was responsible for organizing the defense of Western Europe against the Soviets. The task promised to be a stepping-stone to the US presidency should he be called on, as was widely expected, to be the Republican candidate in 1952. Yet here was Churchill not only accusing him of misrepresentation in his book, but also threatening to embarrass and undermine him in other ways.

By the time Churchill returned to office, how best to defend Western Europe had become a deeply contentious issue. Eisenhower and the Americans were backing a plan that had originated the year before with the French and aimed to deal with fears about a rearmed Germany. The plan, which had come to be known as the European Defense Community (EDC), called for a European army, which would include West Germany, France, Italy, and the Benelux countries. The Germans would not have an independent national army, but rather would be part of an integrated European force. Churchill opposed this plan. It was not that he was against the idea of rearming Germany (he advocated it), nor that he opposed a European army per se (he had called for one in his Strasbourg talk in 1950). He was against the kind of integrated force Eisenhower favored; Churchill preferred a coalition force. In other words, once again the two men were at odds strategically.

Eisenhower was big and broad-shouldered, with a lopsided grin and warm blue eyes. When he was riled, which he seemed to be more often than most men, his face flushed and his eyes went icy. He gnawed the earpiece of his spectacles; he bit his lips with nervous rage. Over breakfast with Eden, he complained about Churchill's threat to air his objections to the current European army proposal in the House of Commons. As Churchill had been an originator of the idea of European unity and indeed of a European army, his active opposition promised to be a major embarrassment, possibly even a deal-breaker.

When Eisenhower made it clear to Eden that he wanted him to get his boss in line, the foreign secretary leaped at the opportunity. After all, in eighteen months, by which time Eden expected Churchill to have left the stage, Eisenhower might well be Prime Minister Eden's American counterpart. Eden suggested that, unlike his boss, he very much liked the current European army plan and that he would do what he could to sway Churchill. He presented himself as the member of the British team on whom the Americans could count not only to listen to reason, but also to intervene on their behalf. As in the 1930s, Eden posed as the sane, sensible alternative to the wayward and erratic Churchill. As in the 1940s, he saw himself and encouraged others to see him as Churchill's restraining hand.

When Eden returned to London, Eisenhower wrote to remind him of the need to get Churchill into a more cooperative mood. Eden's mission proved to be a good deal less difficult than either he or Eisenhower had expected. The prime minister quickly agreed not to attack the EDC proposal in his defense speech in the House of Commons on December 6. Though he liked the European army plan no better than before, he promised that, as long as Britain was not asked to contribute troops, he would make a public statement of his support when he went to Paris later in the month.

It was certainly not like Churchill to yield without a battle royal. The iron refusal to give in that had served him (not to mention the cause of freedom) so well in 1940 had proven utterly exasperating to his American allies when he directed it against them. In his wartime encounters with Eisenhower, Churchill had harangued, pleaded, wept, threatened to resign, and done anything else he thought might help carry his point. On the present occasion, if there was none of the usual Churchillian intransigence, it was not due to any mellowing on his part. Churchill regarded the public statement as a minuscule concession in the interest of keeping the Americans happy prior to his visit, and he was determined to do whatever it took to make sure his trip was a success. Principal among these measures was reassuring both the Germans and the French that they had nothing to fear from his talks with Truman. Before Churchill went to Paris, he conferred with the chancellor of the German Federal Republic, Konrad Adenauer, who was in London at the beginning of December. Churchill hoped to allay German concerns that he and the Americans might be ready to negotiate a deal at Germany's expense.

Before he left for Paris, Churchill made a dramatic surprise move to heal the political wounds in Britain that threatened to weaken his position abroad. The slender Tory majority of seventeen reflected a nation sharply divided after a bruising election campaign. When Churchill gave his speech on defense on December 6 it was widely expected that he would begin by cataloging the weaknesses he had discovered upon taking office. Instead, he made a point not only of complimenting Labour, but also of tearfully praising the very politician, Emanuel Shinwell, who had most savagely and personally criticized him during the general election campaign and whom he had singled out for criticism in his Plymouth speech. As a rule, Churchill believed it was wise to be magnanimous in victory. In the present instance, reaching out to the socialist opposition at the penultimate meeting of Parliament before the winter recess had the distinct advantage that, as a consequence, he would be able to address the Americans as a national leader.

On December 16, Churchill took the night boat train to Paris, where he hoped to ease fears that French interests would not be taken into account in his Washington talks and that France would be excluded when he and the Americans met Stalin at the summit. The French also needed to be reassured that Britain did not oppose the present European army plan.

A cold mist enveloped the Gare du Nord the next morning when Churchill, Eden, and their staffs were met by a large official greeting party. Churchill took his time reviewing the guard of honor of the Garde Républicaine, laboriously inspecting each man's decorations. From the outset, as Eden's private secretary perceived, this was not an easy trip for the heir apparent. The previous month in Paris, Eden had had the stage entirely to himself. Now that he was here with Churchill, he inevitably reverted to playing "second string."

Eden need not have been jealous of all the attention Churchill received on his first trip abroad as peacetime prime minister. Some of it reflected people's fascination with the seventy-seven-year-old leader's physical and mental condition and with how it affected the power dynamic in London. From this point on, everywhere Churchill went, people would be watching, judging, evaluating. There could be no avoiding the questions. How did he look? How did he sound? Was he as sharp as he had been previously? Was he forgetful? Did he make mistakes? Had the mastermind begun to lose his mind? Did he seem overly emotional? Did he fail at times to understand what others said or was it just that he had been unable to hear? How often did he have to leave the room to urinate?

The French ministers who faced Churchill that day whispered among themselves that he had aged "very considerably" during the first eight weeks of his premiership, and predicted that he could not last in office for long. Still, when Churchill dueled with one of his hosts over the latter's assertion that he had been inconsistent in his views on European federation, the others took note that Churchill was still capable of defending himself "vigorously." Would it have been any consolation to Eden to know that the French ministers' scrutinizing eyes had been on the number two man as well? Afterward, at least one minister expressed doubt that, when the time came, Eden would be Churchill's successor.

Having stayed up into the small hours drinking whisky at the British Embassy as he burnished the draft communiqué on his Paris visit, Churchill spent much of the next morning, as was his custom, at work in bed. Eden's follow-up meeting with the French had to be cut off early, as he and Churchill were expected at Supreme Headquarters, where Churchill was to lunch with Eisenhower. Churchill was not happy to have to take time out of his tightly-scheduled two-day visit to drive to Versailles. When Eisenhower had written to propose a meeting which would enable Churchill to see Supreme Headquarters and meet key NATO staff, the prime minister had asked him to lunch at the British embassy instead, as that would be more convenient. Eisenhower fired back a "deliberately grumpy" message to the effect that while he would always go and see his old friend he was disappointed that Churchill did not have the time to visit Supreme Headquarters. A mini-test of wills between them ended with Churchill agreeing to come to Eisenhower on December 18.

Bundled into a heavy woolen overcoat and a felt fedora, Churchill pulled up at Supreme Headquarters twenty minutes late on account of a blinding fog. It was to be his first meeting with Eisenhower since the publication of *Closing the Ring* and Eisenhower was spoiling for a fight. The previous week, after receiving word that Churchill had agreed to come to him, Eisenhower had told seven dinner guests including the US ambassador in France, David Bruce, and the journalist C. L. Sulzberger, that he planned "to give Churchill hell" about his position on the European army. To judge by his remarks to his guests, that was by no means all that was distressing him. Again, his outrage centered on the question of who ought to have listened to whom in 1945. Sulzberger recorded in his diary afterward: "Eisenhower blames Churchill entirely for the political division of Germany which gave Russia such a large share. He said Churchill never had any faith in 'Overlord'—the invasion of Normandy . . . As a result of Churchill's skeptical attitude, unsound political decisions were taken . . . The line was chosen largely because of Churchill's pessimism."

Eisenhower had vowed "to put the bite on Churchill" when he came to Supreme Headquarters. Hardly was Churchill in the door when the fighting began, but it was the prime minister who struck first. Reminding Eisenhower that he had advocated a European army as early as 1950, he went on to explain that it was merely the present inefficient and ineffective plan that he objected to. "I want a fagot of staves bound by a ring of steel," Churchill lectured Eisenhower, "and not a soft putty affair such as is now contemplated." Eisenhower, in turn, "pounded" Churchill, making it clear that he did not think that the prime minister disagreed with the proposal so much as that he just did not understand it. In his diary, Eisenhower judged that Churchill was no longer capable of absorbing new ideas. And there was another problem: "I am back in Europe in a status that is not too greatly different, in his mind, from that which I held with respect to him in World War II," Eisenhower wrote. "To my mind, he simply will not think in terms of today, but rather only those of the war years." At a moment when Eisenhower had his eye on the presidency, being asked to revert to his former relationship with Churchill—who as far as he was concerned ought to have retired anyway—was not a pleasing situation.

Churchill returned to London in anticipation of sailing to America in a little over a week. As his departure drew near, he learned that an announcement of American economic aid to Britain would be made around the time his ship docked in New York. In his recent dealings with the Germans, the French, and the Americans, Churchill had shown that, in certain circumstances at least, he was willing to stoop to conquer. But the idea of arriving in the US in the guise of a "penniless beggar" made him flush with anger. The humiliation, both to his country and to himself, would be too much for this exceedingly proud man to bear.

Churchill believed that during the war he had gone to Washington as an equal and he desperately wanted that to be the case again. Britain's economic weakness was only part of the problem. He deplored the liquidation of its empire begun under Attlee. He bemoaned the loss of influence and the will to rule. His January 1952 trip was very much about recapturing lost dignity by restoring Britain to a prime role in world affairs. In Churchill's view, Britain, "broken and impoverished" though it might be, still had something precious to offer. He hoped to align what he saw as British wisdom and experience with American military might in a concerted effort to win the peace in Europe.

On December 29, hours before he was due to leave on the first leg of his trip, Churchill contacted the American ambassador to say that he desired to speak to him immediately. When Gifford arrived, Churchill interrupted a rare Saturday Cabinet meeting to see him. Weeks after the ambassador had warned against any public call for talks with the Soviets, Churchill had a warning for him. He made it clear that he did not want Britain's financial plight to be thrown in his face as he came off the ship in New York. He demanded that Washington alter the timing of its announcement.

Nothing had been settled when the ambassador left and Churchill returned to the Cabinet Room in foul spirits. By the time he wound up a second late afternoon meeting in anticipation of temporarily entrusting Cabinet affairs to the leadership of Harry Crookshank, he had heard from the American embassy. The offending statement would be postponed until after he came back from the U.S.

At the end of a long, emotionally draining day, Churchill was driven to Waterloo Station, where reporters peppered him with questions as he boarded a special coach bound for Southampton. It was nearly one in the morning when he, Eden (who had left the Foreign Office in Salisbury's care), and the rest of their party of thirty-six boarded the *Queen Mary*. The vessel had docked three days late after the roughest crossing of her captain's nearly forty-year career, but the skipper expected they would probably be able to sail by noon. If all went well, Churchill would arrive in New York on Friday in time for his first scheduled meeting with Truman that afternoon in Washington.

On Sunday morning, Churchill was in bed when the captain came in with a sheepish look on his face. Clearly expecting Churchill to blow up when he heard the news, he informed him that the sixteen-ton port anchor had jammed. The ship could not sail until the anchor was freed. It would likely be another twenty-four hours before they finally headed out to sea. A telegram went off to Truman and the frantic juggling of arrangements began. For the rest of the day, as Cunard Line employees used acetylene torches to blast the anchor loose, Churchill did not budge from his rooms, which had been specially equipped with numerous extra-large ashtrays and a generous stock of brandy.

That evening, Lord Mountbatten, the Fourth Sea Lord, came to dine. In the course of the meal, the prime minister and his outspoken guest clashed over the wisdom of linking Britain's fortunes with that of the Americans. Churchill was not in the mood to hear Mountbatten's anti-American views as he waited to embark on a trip designed to strengthen the Anglo-American bond. In annoyance, he cautioned Mountbatten to avoid expressing any political opinions in the future. "Your one value as a sailor is that you are completely non-political. Take care you remain so!"

The *Queen Mary* finally sailed at noon on December 31. At ten minutes before midnight, the prime minister's entourage joined him for a toast. While his guests listened to Big Ben strike a dozen times on the radio and to the BBC announcer wish listeners a happy new year, the seated Churchill looked as if he were absorbed in thoughts of his own. At last, champagne glass in hand, he hoisted himself up out of his chair. As Churchill swayed with the ship, he gave the impression of being about to give a speech. Instead, he said only, "God save the King," whereupon everyone drank his health.

Before long, the crowd had dwindled to a few intimates. Churchill again seemed preoccupied, and no one spoke until finally he broke the silence. He talked of his trepidation about addressing a joint session of Congress. He worried that the speech, which he had very much wanted to give but had yet to write, might not come off. Later, he returned to the subject when his doctor remained behind to take his pulse, which proved to be rapid and irregular. As Churchill began to undress for bed, he lamented that his mind was no longer as good as it used to be and that these days the prospect of delivering an address—any address—made him uneasy.

Anxiety about his speech to Congress would burden Churchill during much of the trip. In early life, Churchill had slaved to make himself master of the spoken word. Now that he was in his late seventies, however, words no longer came as freely and fluently as they had in the past. Simply to marshal them was an enormous effort. The Washington speech was especially worrisome because of scheduling problems that had required it to be put off until January 17, extending his time on the road to more than three weeks. As it was, Churchill faced several days of difficult talks in Washington before going on to New York and Ottawa; only then would he return for his nationally broadcast congressional appearance and a crucial last session with the president.

On Saturday, January 5, 1952, the *Queen Mary* arrived in New

York in a soaking rain. Churchill flew directly to Washington, where Truman hosted a dinner party in his honor on the presidential yacht. In recent weeks, as Truman prepared to see Churchill, the White House had received various reports on his condition, including one from Ambassador Gifford that stated that he was "definitely aging and . . . no longer able to retain his full clarity and energy for extended periods." In spite of anything those accounts might have led Truman to expect, and in spite of his being in much physical discomfort due to a boil on his leg and a chill contracted at sea, Churchill was pitch-perfect that evening.

When the table had been cleared after dinner, Churchill, Eden, and their group and Truman, Acheson, and theirs returned to the dining saloon for conversation. The president opened the discussion by inviting Churchill to give his American hosts "the benefit of his reflections on the state of the world," with particular reference to the attitude of the Soviet Union.

The theme of Churchill's remarks was fear as the "central factor" in Soviet policy. He suggested that at present the Soviets feared the West's friendship more than they did its enmity. Still, he hoped that an increase in Western strength would encourage Moscow to seek the free world's friendship instead. Churchill lauded Truman's decision to go into Korea, which had led to rearmament in the US and other "freedom-loving countries." By lessening the chances of a Soviet attack, Churchill maintained, rearmament increased the chances of peace. And that was wholly thanks to Truman. In Churchill's telling, the president's actions had been the "turning point" in East-West relations.

Churchill sensed that the president liked to be flattered and deferred to. In his comments, the guest of honor had taken care to do both. He knew that Truman valued brevity, so he limited his talk to five or six minutes. It was no secret that Truman approached Churchill's visit warily. Where Churchill had pushed for informal talks ("so that we can work together easily and intimately"), Truman had demanded an agenda. Truman also had insisted on having his advisers present whenever he met with Churchill. For all the accounts of the aged prime minister's decay, Truman still seemed worried about being outtalked and outflanked.

Truman was not alone in his concern about what Churchill might yet be capable of. The British ambassador to the US, Sir Oliver Franks, had recently reported to London on the "nervousness" among US officials, particularly at the Pentagon, and in parts of the American press that Churchill would dominate the Washington talks "by sheer weight of personality." In his after-dinner remarks, therefore, Churchill did no more than set up the argument he had traveled to America to make. The actual pitch for a meeting with Stalin would come later.

Churchill viewed his first night back in Washington as an unqualified success. At the end of the Potomac cruise, he assured the president that he had never felt relations between their countries to be closer. When Churchill returned to the British embassy after midnight, he rejoiced that he and Truman had been able to talk as equals.

In the formal meetings that followed, Churchill was a good deal less happy with his reception. At times, Truman simply cut Churchill off when he spoke emotionally of the Anglo-American bond. "Thank you, Mr. Prime Minister," Truman would say in a loud, cheerful, patronizing voice that suggested he wanted to keep things moving. "We might pass that on to be worked out by our advisers."

Beforehand, the Americans had determined that any effort to recreate the intimacy of Churchill's old relationship with Roosevelt ought to be discouraged: Churchill must be made to understand that, in 1952, he would be operating under "different circumstances" from those he had experienced in Washington during the war. That attitude did not sit well with Churchill, who bristled at their image of Britain as a junior partner.

One of the principal subjects under discussion in the plenary sessions, whether or not to appoint an American admiral as Supreme Allied Commander of the Atlantic, soon became a focal point for Churchill's resentment. Attlee had been prime minister when an agreement was struck to give the sea command to an American. At the time, Churchill had regarded the arrangement as an insult to Britain and he had been on his feet in the Commons to demand that a British admiral have the job. As the Americans were in command on land, he believed it only fair for the British to lay claim to the seas. Churchill's passion on this subject had not subsided since he became prime minister. If anything, he seemed even more fervent. While he was in Washington, he made it the testing point between the Americans and himself.

Key members of his delegation thought this unwise. At the British embassy on January 7, they urged him not to concentrate on such a purely symbolic issue. To Churchill's horror, Secretary of State for Commonwealth Relations Lord Ismay, First Sea Lord Admiral Sir Rhoderick McGrigor, Chief of the Imperial Staff Field Marshal Sir William Slim and Ambassador Franks—all seemed to regard the appointment as not worth fighting for. The time for the British government to press the case had been when matters were still in Attlee's hands; perhaps then the Americans might have been swayed. As far as Churchill's advisers were concerned, at this late date the only thing to do was acquiesce.

Churchill adamantly disagreed. In spite of what anyone said, he was determined to press Truman to release him from the obligation of the previous government. After Churchill left the room, grumblings were heard that he had lost the ability to see things in perspective. (Churchill's detractors had long claimed that he never had that ability in the first place.)

That day and the next, in the presence of the very advisers who had counseled him to back off, Churchill locked antlers with the Americans. He argued that Britain had earned equality "with British blood" and he insisted that equality was all he was asking for. It was evident to the Americans that he viewed the whole subject of the Atlantic command with a great deal of emotion—rather too much, as far as they were concerned. The Americans perceived something else as well: Churchill's own military staff did not agree with his position.

Churchill's diehard refusal to fall into step on the naval issue created an awkward situation for the British and American teams alike. On one side, his countrymen—most of them, anyway—had no wish to undermine and embarrass him in front of Truman; they just thought he was wrong. On the other side, the Americans acknowledged that Churchill's personal stature required them to secure the new government's pro forma approval even though they had already made a deal with Attlee.

The first round of meetings ended on January 8. At the British embassy that evening, Churchill was so weary that he lay in bed rather than come down to dinner. To be sure, he had managed to hold his ground in the debate about the Atlantic command. He calculated that the Americans would not dare go forward without him. Yet even as Churchill fought to uphold British dignity, he had been at risk of jeopardizing his own.

There had been moments during the talks when Churchill's eyes clouded over alarmingly, moments when he seemed uncertain and confused, and moments when his own team judged that "poor old Winston" was not the most effective advocate for his country's interests. At the fourth plenary session, Churchill had even appeared briefly to forget where he was and to be under the peculiar impression that he was speaking to the British Cabinet. Afterward, Eden's private secretary noted in his diary that the Americans had "bravely and loyally" extricated the British delegation from this predicament.

Tonight, when Moran came in to see him, Churchill complained that he had been kicked about so much since he left London that he could hardly any longer tell night from day. At his age, Churchill explained, the problem was not attending meetings as much as thinking out what he wanted to say.

While Churchill conferred with his doctor, Eden was downstairs airing some complaints of his own. Due to return to Europe directly from Canada, the number two man was distressed about the wording of the press communiqué to be issued before they left Washington. Like his boss, Eden harped on the theme of equality, but with a somewhat different emphasis. He noted that every paragraph began, "The Prime Minister . . ." or "The Foreign Secretary was instructed . . ." "No one instructs me," Eden railed. "The Prime Minister and I are colleagues."

Churchill spent January 9–16 in New York and Canada. Back in Washington on the 17th, he was still at work on his speech when Deputy Under Secretary of State Sir Roger Makins came in to announce that a car was ready to take him to the Capitol, where he was to go on in about forty minutes. Churchill, who had yet to rouse himself from bed, was absorbed in puzzling over a passage about British participation in the Korean War.

"If the Chinese cross the Yalu River, our reply will be—what?" he asked.

"Prompt, resolute and effective," Makins shot back.

"Excellent," said Churchill. He inserted that phrase in the body of his speech. Then he dressed hurriedly and, assisted by a motorcycle escort, managed to get to the House of Representatives two minutes early.

In the course of a thirty-seven-minute address that was carried by all the radio and television networks, Churchill evoked a changed world in which former allies had become foes, former foes had become allies, conquered countries had been liberated, and liberated countries had been enslaved by Communism. Speaking in a firm, strong voice and rarely hesitating over his phrases, Churchill revisited the theme he had developed privately that first night on the presidential yacht. He argued that the consequences of Truman's actions in Korea extended far beyond that region. The "vast process of American rearmament" had already altered the international balance of power and might well, "if we all persevere steadfastly and loyally together, avert the danger of a third world war."

A satisfied smile erupted on Churchill's face as he walked slowly but energetically through the standing, cheering crowd. When he reached his car, he sank back into the seat, visibly relieved.

He could not relax for long. The following afternoon, Friday,

January 18, he and Truman were scheduled to discuss Soviet relations. Now that he had further set up his argument in Congress, Churchill was finally about to make his case for a return to the table with Stalin. It would be but a short distance from crediting Truman with having shifted the balance of power, to urging him to capitalize on that achievement by securing the postwar settlement.

On Friday morning, Churchill, Admiral McGrigor and other military staff visited the Pentagon, where, at Churchill's request, defense secretary Robert Lovett briefed them on American war plans. Afterward, Churchill went back to the embassy for a nap. The rest of the British contingent remained at the Pentagon, where, in conjunction with American officials, they drafted a joint communiqué that ratified previous decisions about the Atlantic command, including the agreement to name an American to the top post.

Shortly before three that afternoon, Churchill was shown into the president's anteroom at the White House. He was joined there by Admiral McGrigor and the other military men, who surprised him with a copy of the joint communiqué. Minutes before his climactic session with Truman, Churchill was confronted with the fact that the very advisers whom he had previously defied had now chosen to stand up to him. In full view of the Americans, his military staff had dared to take a position contrary to the one he espoused.

Because of his staff's participation, the document was no longer just about Britain's decline—it was about Churchill's. The weakness of his leadership was on display for all to see. When he finished reading, his mute comment was to rip up the text and throw the bits into the air like confetti.

"Hurricane warnings along the Potomac," said Admiral McGrigor, as the noticeably "shaken" British delegation entered the Cabinet Room, where the Americans, with the exception of Truman, had already assembled. In the moments before Truman appeared, Churchill lingered in the anteroom to improvise a draft of his own.

When both leaders had taken their seats, the president opened the proceedings. He began by saying that it was regrettable that all good things must come to an end and that that included Churchill's visit. He insisted he would be sorry to see the prime minister leave, and went on to suggest that Churchill use the present occasion to speak about whatever else was on his mind. For many weeks, the White House had been bracing for Churchill to make his case for a meeting with Stalin. Now, the president had given Churchill his cue. Truman had offered him the perfect opportunity to lay out his thoughts on East-West relations (the day's official topic, after all). That Truman was well-armed with previously-prepared arguments against a summit meeting, and that he was ready and eager to shoot down Churchill's proposals, would have come as no surprise to the prime minister. Churchill had not, however, expected to be ambushed by the members of his own team.

Hot with indignation, he began by addressing the lack of agreement on the naval issue. If Britain and the US could not come to terms, he warned, he would have to take up the matter with NATO, and the rift between the two nations would be made public. That was something neither side wanted. He delivered a "majestic" address on Britain's proud history as a ruler of the seas. He asked the Americans, in the plenitude of their power, to allow his country to continue to play its traditional role "upon that western sea whose floor is white with the bones of Englishmen." Dean Acheson later went so far as to call the performance "one of Mr. Churchill's greatest speeches." It was good enough, at any rate, to impel Sir Oliver Franks to pass a note to warn his friend Acheson to be "very, very careful"—that is, not to allow himself to be dazzled.

Churchill made it clear that he could not accept the joint communiqué that had been prepared in his absence. In its place he had written out "a little statement" of his own, which he proceeded to read aloud. It was a measure of Churchill's excruciating isolation in the room that it fell to Franks, who opposed him on this, to hand the prime minister's draft to Truman. After some discussion, Truman asked Acheson and Secretary of Defense Lovett to study the new document. Churchill's powerful presentation had sent the meeting completely off the rails. Suddenly, everyone was waiting to hear from Acheson and Lovett.

In this unsettled atmosphere, Churchill finally got around to raising the possibility of a conference with the Soviet Union. That topic was supposed to have been the centerpiece not just of the fifth plenary session but also of his entire trip. Now, it came across almost as an afterthought, a side issue. On various occasions in the past, Churchill had spoken beautifully and affectingly of his desire to return to the summit. Today, however, he had used himself up on the naval question. He had squandered his dwindling energies and capacities on a lost cause.

Truman easily brushed aside Churchill's halting comments. The fact that Moscow had recently referred to a top US general in Korea as a cannibal suggested to Truman that the Soviets were not in a conciliatory mood and that a top-level meeting would not be helpful.

Churchill had been looking forward to this conversation for months, yet instead of responding to Truman's argument he abruptly changed the subject and asked how Acheson and Lovett were doing. At length, Acheson proposed that he might be able to suggest a compromise. He, Lovett, McGrigor, Franks, Air Marshal Sir William Elliot, and some of the others went off together to the president's office to draft a new document.

While Churchill waited in the Cabinet Room for a good thirty minutes, it was as if all the juice had seeped out of him. Now that the tired titan no longer posed a threat, Truman magnanimously gave him several openings to at least complete his summit pitch, but Churchill passed up every one. Eventually, Churchill sent his private secretary, Jock Colville, to see how much longer Acheson and the others planned to take. This prompted Truman to protest that he was enjoying their talk and was in no hurry. Churchill, however, was eager for the painful and increasingly pointless encounter to end. He had failed to make his case about the Soviets. When the drafting group returned and Acheson read the revised communiqué aloud, it was evident that Churchill had been unsuccessful in this area as well. The new document still provided for the appointment of an American admiral as supreme commander.

Churchill's antagonists on both sides of the table waited an "interminable minute" for his response.

Unwilling to prolong this he said finally, "I accept."

XIII

NAKED AMONG MINE ENEMIES

New York, January 1952

On the morning of Saturday, January 19, 1952, Churchill spent much of the train trip to New York gazing out the window. Ordinarily, he preferred to work, play cards, or devour a stack of newspapers, but today he was not in the mood. He made passing comments to his doctor and others in his traveling party of fourteen. Yet he seemed too weary and distracted to focus on any one topic for long. At length, he disappeared to his compartment, where he slept deeply for an hour. Even that did not seem to help, and as the train drew into the station Churchill complained that he was still very tired. Before he sailed he intended to spend forty-eight hours at the home of his friend Bernard Baruch, where he hoped to recover. The eighty-one-year-old American financier—who liked to say that as far as he was concerned "old age" was always fifteen years older than he was—arranged to have Churchill collected from Pennsylvania Station so that he could avoid a barrage of reporters' questions.

Far from reviving in New York, Churchill fell ill as the muggy weather turned icy overnight. "Such is life nowadays," he reflected, when a head cold required him to disappoint a great many people by failing to appear in a ticker-tape parade and a ceremony in his honor at City Hall. Not feeling well enough to rise from his bed, a blue-pajama-clad Churchill received the archbishop of New York and the mayor at the Baruch residence. He leaned forward slightly from a stack of pillows as Mayor Impellitteri placed the city's Medal of Honor round his neck. Late that same day, Tuesday the 22nd, Churchill emerged onto Fifth Avenue and Sixty-sixth Street wearing a gray storm coat with a fur collar, en route to the *Queen Mary*, which was due to sail at midnight.

During much of the voyage, Churchill lay in his suite on the main deck, where late one morning he dreamed that he was so sick that he could neither see nor walk straight. Awakening with a jolt, he hurried out of bed to test his ability to walk, which proved to be unimpaired.

Two days later, on Monday, January 28, Clementine Churchill was waiting at Southampton when the *Queen Mary* docked. He looked forward to dining alone with her. But when the couple reached Waterloo Station they were met by Eden, Butler, Crookshank, and others whose business could not be delayed. The ministers accompanied Churchill home to lay out the crisis that faced him the next day when Parliament met again after a recess of more than seven weeks.

The congressional address that had caused Churchill such anxiety beforehand had set off a political earthquake at home. Of all things, it was the four words that Sir Roger Makins had casually suggested be added to the text that had produced the most severe tremors. When Makins proposed the phrase "prompt, resolute and effective," he had merely been trying to rush Churchill out of bed and get him to Congress on time. Churchill had liked the sound of the phrase, to which he had attached no further significance than that Britain and the US would respond firmly to any breach of a truce in Korea. The Labour opposition, however, had since read a great deal more into it. They interpreted it to mean that Churchill had secretly agreed to an American plan to extend the Korean War to Communist China, to bomb Chinese cities, and even possibly to employ an atomic bomb. They used it to revive the charges of warmongering that had stung Churchill in the general election. In recent days, concern about Churchill's choice of words had taken root in a broad section of the press, and he was coming home to demands from various quarters, not just the left, that he immediately clarify what he had agreed to in his talks with Truman.

Churchill hoped to forestall a full-dress account for at least several days. He had recovered from his head cold but had yet to regain his strength after the trip and was not ready to take on anything like this just yet. The next morning, he told the Cabinet of his intention to reply to the four prime minister's questions that he had for that day's session. Then, he would invite the House to await a complete statement to be made at the start of the foreign affairs debate the following week.

Eden said he might "get away" with it, but went on to remind Churchill that there was a campaign against him and that the House of Commons was "very anxious" to know exactly what guarantees he had given to Truman. Even after the Cabinet meeting had been interrupted by a message from the Labour whip asking for a statement that day, Churchill was still inclined to stall.

"Let's see what fuss they make today," Churchill said. "If there's great trouble, I will promise to make it tomorrow."

"Then you will have it forced out of you," warned the persnickety Conservative chief whip, Patrick Buchan-Hepburn. "Is that good?" Churchill did not seem to think it would be a problem.

At half past two that afternoon, Parliament met for the first time since before Christmas. Back in December, Churchill had offered the socialists a peace pipe. Today, he pretended not to notice they were all wearing war paint. Asked for a full accounting of his Washington visit, he replied that he had thought it would be "more convenient" for the House if he dealt with the matter in opening the debate on foreign affairs which was set to begin the following Tuesday.

"As, however, I learn it is the wish of the Opposition that some statement should be made," Churchill continued, "I should be prepared to make a preliminary statement tomorrow."

Attlee reminded him in the tone of a stern and disapproving schoolmaster that there was considerable anxiety about certain matters that had occurred overseas. Churchill retorted, "It is on those points on which there is considerable anxiety that I may, in comparatively few sentences, endeavor to enlighten the House and, I trust, relieve the anxieties which prevail."

Back in Parliament the next day, Churchill deadpanned that he had used the phrase "prompt, resolute and effective" because he preferred those words to "tardy, timid and fatuous." In fact, he insisted, he had carefully chosen his language to make it plain to the Communists that Britain and the US would "work together in true comradeship" in the event of a renewal of hostilities in Korea. He stressed that that was all he had meant and that there was no sense in which he had been signaling a shift in Far East policy. Churchill's explanation, not to mention his sometimes insouciant manner, drew a good deal of opposition scorn.

So did his decision to send Eden to open the foreign affairs debate on February 5, though Labour had had every reason to expect Churchill to speak first. Eden, when he faced the House, did not seem especially pleased with his assignment either. He was visibly "irritated" when Churchill persisted in conversing with Buchan-Hepburn while Eden struggled to defend him. Labour's surprise response to Eden's efforts was to harpoon Churchill with a personal censure motion. The artfully worded motion endorsed the government's Far East policy as stated by Eden, but regretted Churchill's "failure to give adequate expression to this policy in the course of his recent visit to the United States."

The claim infuriated Churchill, who set to work preparing a real fighting speech in response to what he viewed as a "shameful" personal attack. As far as he was concerned, it was both absurd and unjust. He had gone to America in quest of peace, yet he stood accused of having set his country on a path to war. He had defied the Americans in the name of equality, yet the opposition assailed him for having allowed Britain to drift into subservience.

On the morning of Wednesday, February 6, Churchill rose early in anticipation of addressing the House that afternoon. He was at work in bed with all his papers around him and a candle for his cigar when Edward Ford, the King's assistant private secretary, was unexpectedly shown in.

"I've got bad news," the visitor said. "The King died this morning."

"Bad news?" said Churchill. "The worst!" As he cast aside the pages he had been absorbed in only a moment before, he remarked, "How unimportant these matters seem."

After Ford left, Jock Colville entered the bedroom, where the prime minister sat looking straight ahead. Tears streamed down Churchill's face as he reminisced about his wartime comradeship with George VI. Colville, who also had served as Princess Elizabeth's private secretary, tried to reassure him by saying how well he would get on with the new Queen, but Churchill was inconsolable.

"I hardly know her," Churchill despaired, "and she is only a child."

At half past eleven, he opened a special Cabinet meeting by announcing that the King had died in his sleep at Sandringham the previous night. Then he turned to the enormous volume of business that suddenly confronted the government. One week previously, Churchill had accompanied the King and Queen, as well as Philip's uncle and aunt Lord and Lady Mountbatten, to London Airport to see Princess Elizabeth and her husband off on the Commonwealth tour she was making in place of her ailing father. She was still in Kenya, its first stop, at the time of his death and would have to come home at once, but the thought of her in an airplane filled Churchill with dread. Initially, it seemed it might be best if the Cabinet recommended that she return by sea. But the trip would take fourteen days, and even as he contemplated the idea, he doubted that they would be able to prevent her from boarding a plane immediately. By the time the Cabinet met again in a few hours, Churchill was able to report that the royal party was already en route and expected the next day.

That night, the House of Commons was filled to overflowing as Churchill slowly rose from his seat and, in a low, soulful voice, took the oath of allegiance to the new sovereign. Then he signed the roll and left the chamber.

Among themselves a number of opposition members admitted to being "terribly disappointed" that the death of King George had postponed the fierce high-stakes debate that otherwise would have played out that day in the House. Churchill's wartime scourge Aneurin Bevan, whose gifts of oratory were thought by some to be on a par with the prime minister's, had been pumped up to lead the attack. There was upset in the Bevan camp that at the last minute Churchill had been rescued by events, and there was concern that Labour would have a hard time recapturing the momentum against the prime minister. "No doubt we shall revive the issue," Richard Crossman, a Bevanite, confided to his diary on February 6, "but, in the flood of sentimentality which Royal deaths and innocent young couples coming to the Throne create, we shall have to bide our time."

Crossman acknowledged that the parliamentarians' "hard-boiled" reaction—his own included—to the King's demise was out of sync with what most Britons appeared to be feeling at the time. "Directly you got outside, you certainly realized that the newspapers were not sentimentalizing when they described the nation's feeling of personal loss."

On the morning of February 7, it fell to Eden to preside over the Cabinet. Churchill had not awakened in time—"being much tired with the emotion of yesterday." Hardly had Churchill cast aside his fighting speech when he had to pivot to the preparation of the first of two eulogies of George VI. Originally scheduled for Wednesday night, his nationwide address had been moved to Thursday at 9 p.m., to take place after the royal party had safely returned.

In the car to London Airport, Churchill wept as he dictated portions of that night's radio broadcast to a secretary. At the airport, where he stood on the tarmac with Eden, Salisbury, Attlee, Lord and Lady Mountbatten and other dignitaries, the seventy-seven-year-old prime minister gave a low bow as the black-clad twenty-five-year-old Queen emerged from the aircraft. Churchill was the first to welcome her, but he was too emotional to speak.

Churchill resumed dictating his eulogy as he was driven back to London. By the time he reached Number Ten, the text of the address was finished, but the effort of composition had exhausted him. While he rested, staff members were told he was still at work.

When Churchill went on the air that evening, he managed to accomplish something he had often done to great effect during the war: he connected with his audience on a visceral level. He gave voice to the feelings of the millions of Britons who had watched with anxiety—in newspaper photographs and more recently in television images on the day of Elizabeth's departure for Kenya—as George VI publicly neared the close of his life. The King, Churchill said, had walked bravely with death as if death had been a companion. "In the end, death came as a friend; and, after a happy day of sunshine and sport, and after 'good night' to those who loved him best, he fell asleep as every man or woman who strives to fear God and nothing else in the world, may hope to do."

The emotion was deeply felt, but the calculating, political part of Churchill's brain had not stopped churning either. He moved seamlessly from evoking the past to contemplating, not to mention staking a passionate personal claim on, what was to come. For the first time the aged premier hitched his wagon to the future of the young monarch. He noted that some of the greatest periods of Britain's history had unfolded under the rule of its queens. "I, whose youth was passed in the august, unchallenged and tranquil glories of the Victorian era may well feel a thrill in invoking once more the prayer and the anthem, 'God Save the Queen.'" The association was significant, it having long been Churchill's dream to secure an enduring peace such as Britain had last enjoyed in his Victorian youth.

By the time Churchill addressed the House of Commons four days later, he had managed to insert both Elizabeth and her late father into the grand narrative of his second premiership. He spoke of the end of the war, when the surmounting of one menace had seemed only to be succeeded by the shadow of another. He recalled George VI's disappointment that victory had led neither to real peace nor security. And he expressed a wish that the accession of Elizabeth II might coincide with the beginning of a new era of true and lasting peace. Thus Churchill sought to connect the Queen to his goal of a parley at the summit.

Soon he was linking her to the question of his retirement as well. Churchill was pointedly in no rush to see Elizabeth crowned. After he had delivered his parliamentary eulogy, he discussed the date of the coronation with the Cabinet. On the argument that the economic crisis ruled out 1952, he pressed for a distant date. He maintained that were the coronation to take place the following year, it would have a "steadying effect" on the country. The Cabinet concurred, and on February 12 he advised the Queen that the ceremony ought to be staged in the spring of 1953. After that had been settled, he set himself a new timelock. He began saying that he intended to remain in office—until the coronation.

Within a matter of days, Churchill had come a long way: from despairing about the accession of a young Queen to finding her supremely useful to his purposes. He was inclined to adopt a paternal attitude in his personal relations with her. Some observers saw his relationship with Elizabeth as a reprise of the formative role that Lord Melbourne had played with the young and malleable Queen Victoria. But there were crucial differences between Elizabeth and her great-great-grandmother at the outset of their reigns. For one thing, Elizabeth was a married woman. Though her husband's uncle, who had been like a father to him, was known to have had a hand in engineering the marriage, it was also the case that the union of Elizabeth and Philip was "a real love-match." "I've seen the man I'm going to marry!" the thirteen-year-old Princess Elizabeth had told her sister Princess Margaret after she met their eighteen-year-old cousin in 1939.

The question of how much power Philip would exert was an important one. Churchill was also very concerned about what Mountbatten's role would be. (Macmillan was not alone in his suspicions that Mountbatten aimed to be "the power behind the Throne.") Churchill and Mountbatten had collided over India, and lately over Mountbatten's anti-American opinions. At a moment when Churchill was still waiting to get Uncle Sam behind his Soviet initiative, could he afford to have "Uncle Dickie" in the mix?

In anticipation of the state funeral on February 15, Mountbatten turned up at the Earl Marshal's meeting, where he proposed that Norwegian and Danish sailors take part in the ceremonies. When this was reported at a Cabinet session, Churchill grumbled, "Why was he present?" "As Fourth Sea Lord, I suppose," said Eden, "and also as a member of the Family." After the funeral, Churchill's annoyance escalated to alarm when word reached him that Mountbatten had been heard to say that "the House of Mountbatten now reigned." Churchill learned of the episode on February 18 from Jock Colville, who had heard it from George VI's mother, Queen Mary (no friend or admirer of Churchill's, by the way), who in turn had heard it from Prince Ernst August of Hanover, who had been present at the royal house party where Mountbatten committed the indiscretion.

Immediately Churchill repeated the story to his Cabinet. Thirty-five years before, the family name Windsor had been adopted by a decree of King George V. Now, Churchill spoke of both Queen Mary's and his own vehement opposition to a name change. The ministers agreed that Windsor should be retained, and they invited the prime minister to communicate their views to Her Majesty. The encounter, set to take place the following evening, February 19, was going to require a delicate touch. However Churchill cast it, he was about to have to ask her in effect to distance herself from, and almost certainly to embarrass, her husband by participating in a rebuff to his family's ambitions.

All this was taking place as the moratorium on partisan strife drew to a close. After the King's burial, Churchill had spent the weekend at Chartwell, where he resumed work on his answer to the personal censure motion that had been hanging over him for days. Back in London on February 18, he sent a draft to Dean Acheson, who had come for the funeral. Would the US secretary of state review Churchill's efforts to be certain he had revealed nothing to which Washington might object? To Acheson's eye, the speech did indeed contain top-secret information and he promptly wrote with his objections.

The next day, Acheson, who had just been to see the Queen about unrelated matters, was shown into the Cabinet Room. Churchill, who had yet to see her himself about the Mountbatten affair, barely welcomed him. Bent forlornly over the baize-covered table, the prime minister muttered not sentences but phrases. Acheson heard him say "naked among mine enemies" and "the sword stricken from my hand." Determined to ignore Churchill's gloom, the visitor genially inquired how he might be of service. Churchill claimed that Acheson had vetoed the heart of his speech and asked whether he could suggest a substitute. Though Acheson had come prepared with an alternative, he protested that to revise Churchill's prose would be impertinent. Churchill waved this aside and, at length, his spirits lifted materially as he listened to Acheson's version—twice. Declaring "I can win with that," he seized the paper from Acheson's hand and read through it himself with pleasure.

Churchill also got what he wanted out of his audience with the Queen. Afterward, he was able to report to the Cabinet that Elizabeth had agreed that she and her descendants would bear the name of Windsor. The battle was far from over, for the consort himself had yet to have his say, but for the moment Churchill was able to put the issue aside.

At a Cabinet meeting on Thursday, February 21, he made it clear that he needed to focus on his appearances in the House of Commons the following week. The censure debate was set for February 26, and two days after that he was to open the defense debate. Eden disclosed to Acheson that the Tories were allowing Churchill to fret about the charges against him, though they were confident the Labour motion would easily be defeated. Churchill knew he had the numbers as well, but it was not enough to beat his accusers; he wanted to decimate them.

Late Thursday afternoon, Churchill went upstairs to his bedroom at Number Ten for a nap. When he awoke, it was about half past six. He picked up the telephone, but when he tried to speak to the operator he could not think of the words he wanted. Other, useless words filled his thoughts. As he feared he knew what was happening to him, he remained silent.

XIV

I LIVE HERE, DON'T I?

London, 1952

After twenty-five days of running on empty since he returned from America, Churchill had suffered another arterial spasm. He let about an hour pass, and when he was able to speak again he summoned his doctor. Seated on the edge of his bed when Moran came in, Churchill recounted the telephone incident and asked what it meant. Was he going to have another stroke? And what were the chances the aphasia would recur? Moran explained that the blood flow to the speech center had been diminished, and that as a result Churchill had been temporarily unable to find the words he wanted. By the doctor's reckoning, Churchill had two options if he wished to avoid another stroke. Either he had to retire immediately, or he needed at least to arrange matters so that he would be under less strain.

What, Moran asked, did Churchill wish him to do if he ran into Clemmie on the way out? Churchill said she was in bed with a cold and he proposed they go to her room to tell her what had happened.

Had it been up to Clementine Churchill, her husband would never have become prime minister again. He had already had one stroke, and it was her nightmare that someday he would break down during a speech, discover he was no longer capable of conducting affairs, and end up like Lord Randolph Churchill "the principal mourner at his own protracted funeral." Nevertheless, rather than upset him further, she listened calmly and silently to his and the doctor's account. Throughout, her demeanor was grave but composed. She judged that this was not the time to admonish, but rather to envelop her husband with care and concern.

On the assumption that not only would Churchill refuse to step aside but that the sudden loss of power could actually prove detrimental to his health, Moran acted immediately to try to give him a bit more time in office. Of course, the arterial spasm might be the harbinger of an imminent full stroke. In the event Churchill survived, he would have little choice but to hand over. If, however, a clot did not form in the next few days, major decisions would have to be taken. The episode had been a warning that unless Churchill agreed to husband his strength, there would likely be severe consequences in six months or less.

Moran, for his part, had no idea whether a prime minister's burden could be eased. He calculated that it would be best to ask someone in the Cabinet. At the same time, he was adamant that he did not want to put a weapon in the form of confidential information in the hands of any individual who might later use it against his patient. He therefore decided to confide in the one minister renowned for his objectivity and complete absence of personal ambition. On Friday morning February 22, Moran asked Jock Colville to arrange a meeting with Salisbury.

Within two hours, Moran and Colville were ushered into Salisbury's office. After the doctor revealed what had happened the night before, he asked whether Churchill's duties could be cut. Salisbury replied that a prime minister could not simply cast off his responsibilities. The solution, he suggested, might be for Churchill to continue as prime minister from the House of Lords, leaving Eden to take over for him in the Commons. This unusual arrangement would allow Churchill to realize his wish to stay in office until after the coronation, and in the meantime it would make Eden prime minister in all but name.

The others liked the idea, but the inevitable question arose: who would be the best emissary to put it to Churchill? Lord Cherwell (Churchill's friend) and Christopher Soames (the husband of Churchill's daughter Mary) were mentioned. The prime minister adored both men, but whether either would be able to persuade him was questionable. Finally, Moran suggested that there was only one person who would be able to convince Churchill. Three days after the prime minister had been in to see Queen Elizabeth about the royal family name, it was agreed that they would appeal to her to talk to him.

That afternoon at Buckingham Palace, Sir Alan Lascelles, now the Queen's private secretary, concurred that something needed to be done. Lascelles had lately noted the uneven quality of the prime minister's performance. There were days when he seemed fine and Lascelles asked himself why he had been worrying; there were also days when Churchill seemed incapable of grasping the point of a discussion. Even so, Lascelles refused to involve the Queen. He informed Moran and Colville that if she did speak to Churchill, he no doubt would thank her graciously for her suggestion before politely dismissing it. She was simply too young and inexperienced to have an effect. Under the circumstances, Lascelles thought Moran should be the one to approach him. The physician disagreed. He argued that Churchill would listen to him solely on medical matters. Moran was ready to warn his patient that he must not continue at his present pace, but someone else would have to broach Salisbury's plan.

At any rate, nothing could even be attempted until after the foreign affairs, defense, and finance debates. To confront Churchill beforehand would only multiply his troubles at a moment when he needed to cut back. Besides, faced with a personal censure motion, Churchill surely would not consider walking out at this point, lest it be said that he had run from a challenge. For the moment, their little group could only hope that he made it safely through the next ten days.

Deeply concerned about his boss, Colville sought to ensure that Churchill did not overtax himself. That weekend at Chartwell, where Churchill had gone to finish his reply to his accusers,

Colville urged him to let someone else assume the burden of the defense speech. Churchill's antennae tended to be hypersensitive to intrigues of any kind, and the private secretary's suggestion that he delegate the speech to Secretary of State for War Anthony Head fired his suspicions. As far as Churchill was aware, only his wife and doctor knew about the arterial spasm. He had no knowledge that Colville had been informed, let alone Salisbury or Lascelles. Abruptly, he turned on Colville and demanded to know whether he had been talking to Moran. Colville insisted he had not.

Instead of delegating the defense speech, Churchill postponed it until March 5 and soon he rescheduled the finance debate to take place on the 11th. As a consequence, the critical ten days of waiting and worrying stretched to nearly three weeks.

Those of Churchill's colleagues who knew about the arterial spasm were not alone in their concern that he might not be up to the exertions that faced him. On the afternoon of the renewed censure debate, Churchill himself questioned whether he would be able to make a speech at all. Characteristically, he had lavished infinite care on the preparation of his address, but now wondered whether he would have the stamina to bring it off. The early signs were not promising. To his alarm, he was dull and listless at lunch and he continued to feel that way after he took his place in the House. For sixty-five minutes he sat through a "cumbrous, ill-at-ease" speech by the former Labour foreign secretary Herbert Morrison that not a few listeners found interminable. It enraged Churchill to hear his trip to Washington dismissed as a doubtful mission by the wrong man at the wrong time, but indignation had yet to animate him. Before he rose to reply he noted that he was still shaky and fatigued.

Churchill warmed to the cheers that greeted him and as he began to speak he felt his strength increase. "He is looking white and fatty," observed the young Tory member Nigel Nicolson, "a most unhealthy look, you would say if he were anyone else, but somehow out of this sickly mountain comes a volcanic flash." Churchill's speech was a tour de force. He exposed his critics as hypocrites who, when they were in power, had done the very things he now stood accused of. Assailed as a warmonger, he disclosed that the Attlee government had been secretly constructing an atomic bomb and had erected "at the expense of many scores of millions of pounds" a facility for its regular production. Attacked for having made unprecedented commitments in Washington, he revealed that his accusers had themselves previously pledged to support US bombing across the Yalu River in the event of heavy air assaults from Chinese bases. He tartly added that he himself had yet to go so far. Pandemonium erupted in the chamber. The Labour front bench "sat stunned and white."

As he had been known to do, Churchill cunningly set his enemies against one another so they would have less strength left to go at him. Among his tasty revelations was that Attlee, Morrison, and their group had kept both secrets not just from the Conservatives but also from Labour's own left wing. The latter faction, headed by Aneurin Bevan, was at the time vying for control of the party. Churchill's disclosures provided the Bevanites with ammunition against the more moderate Labour leadership. Privately, Churchill would not have minded if the radicals, whose ideas and leader were anathema to him, won out over Attlee. He calculated that their extreme views would alienate a good many middle-class voters who had been drawn to the Labour banner since 1945.

Conservative cheers and sarcastic Labour cries threatened to drown out Macmillan when he jauntily declared that the censure motion had backfired and urged the House to reject it with the scorn it merited. After the motion was defeated, Moran advised his patient to go home. Too worked up to leave, Churchill bustled to the Smoking Room to learn what his colleagues had thought of his performance.

A newly confident Churchill returned to the charge at the defense debate the following week. He twitted Labour about the cleavage in their ranks. The angrier they became the merrier he seemed. At one point, he managed to get so far under Emanuel Shinwell's skin that the former defense minister burst out, "I wish Mr. Churchill would stop interrupting in an inaudible fashion. He sits there muttering. He is smirking and grinning all over his face as if he is enjoying himself hugely." Never was the premier's pleasure more manifest than when the Bevanites twice openly defied Attlee's voting orders.

Meanwhile, internecine warfare had also erupted in Churchill's Cabinet. In anticipation of the finance debate, Eden had targeted Butler's plan to float the pound. Macmillan, sensing that the plan posed a threat to his housing program, formed a tag team with Eden. They took turns pummeling Butler, who tended to shrink from head-on clashes. Churchill, having initially supported the plan, stood aloof for the most part, but at length he sided with the attackers. Within the Cabinet, the chancellor's defeat registered as a significant political victory for Eden.

The maneuvers, however, were far from finished. When Butler lost out to Eden, he had yet to present his maiden budget in the House of Commons. After consulting with Churchill on March 8, he produced a speech that was received triumphantly at the finance debate. As cheering MPs rose in their places to wave their order papers, the prime minister, near tears, offered Butler the highest form of praise. He compared him to his own father, Lord Randolph Churchill. In spite of what had gone on behind closed doors, the strength of Butler's parliamentary performance sparked a round of suggestions in the press that he had taken the lead over Eden in the race to succeed Churchill.

After the finance debate, Moran delivered his medical ultimatum in the form of a letter, a copy of which he sent to Clementine Churchill. In recent days, Churchill had spoken of being prepared to die in harness. Moran pointed out to Mrs. Churchill, as if she did not know it already, that her husband was far from putting the whole case when he said that. What one dreaded much more, the physician explained, stoking Clementine's deepest fears, was an attack which would leave him permanently disabled. This was what they were striving to avoid.

Churchill was in no mood to be ordered to work less, certainly not after his latest triumphs. He brusquely dismissed the letter and when Clementine took it on herself to propose Salisbury's plan, he called the arrangement impractical and refused even to consider it. Churchill conceded to Moran that he was winding down physically and mentally. Nevertheless, he advised the doctor to stop worrying and reminded him that everybody has to die at some point.

Stalin's personal physician, Vladimir Vinogradov, did not get off so easily when his patient rejected his advice. Furious at being told to retire lest he suffer a severe stroke, Stalin destroyed his medical records and vowed to stay away from doctors in the future. He later went so far as to have Vinogradov arrested and tortured for being part of a Western-inspired plot by eminent doctors to murder the nation's leaders through incorrect diagnosis and faulty treatment. As it happened, on the day Moran sent his note to Churchill, the prime minister was conferring with his Cabinet about a message from the ailing Stalin that had been presented to the British, American, and French ambassadors in Moscow. Reacting to Western efforts to establish a continental defense force and to rearm the Germans, Stalin proposed a four-power conference, possibly at the highest level, on the creation of a reunified neutral Germany. Churchill recognized that the offer might simply be a propaganda ploy and he saw that a neutral Germany, were it to be undermined by the Soviets, might soon go the way of Czechoslovakia. Still, another chance to face Stalin was exactly what he had been longing to arrange.

Despite this, from the moment Churchill returned to office, he had also been operating on the premise that serious settlement talks could not occur until after the US presidential elections. Although this timetable did not suit him, he remained intent on acting in tandem with the Americans, which meant he was going to have to resist Stalin's overture and wait another six months at least.

Could Churchill last? At times, he had his doubts. He had bested Labour in recent debates, but now a contingent of about fifty young socialists in the House took to heckling him regularly about his age and growing deafness. They hooted at his entrances and exits, jeered his pronouncements, and filled the chamber with derisive laughter when they caught him having to lean forward and cup his hand to his ear in order to follow the give-and-take.

It was bad enough that the opposition openly and repeatedly questioned his competence. Before long, Conservatives were raising questions of their own and Churchill became a piñata for both sides. Young Tories in particular were unnerved by heavy Conservative losses in the county council and borough elections in spring 1952. Though Churchill had warned at the start of his peacetime government that the economic recovery would not take place overnight, Britons were naturally impatient after seven years of postwar austerity. At a moment when jubilant Labour spokesmen were already calling for a new general election, Conservatives began to wonder whether younger, more effective party leadership might be in order. On April 9, after being mocked and taunted by his young Labour persecutors in the House, Churchill went on to assure his young Conservative critics in the 1922 Committee that things would get better—in time. He also pledged not to cling to power if he began to fail physically or mentally.

Following a bout of bronchitis that kept him in bed for a week and excited much concern about his health, Churchill returned to Parliament with a hearing aid in his left ear. By turns, he insisted his health was better than it had been in a long time and complained that his zest was diminished. He boasted that he could still put his enemies on their backs but lamented that his "old brain" did not work as well as it used to. He told Eden that he looked forward to working with the next administration in Washington but suggested (for the umpteenth time) that it would not be long before he handed over. As far as the hearing aid was concerned, he found it uncomfortable and frequently tore it out in disgust.

Three months after Churchill had declined to go to the House of Lords, Salisbury, Crookshank, Stuart, and Buchan-Hepburn met to formulate a new ultimatum. Amid a blizzard of press rumors that Churchill might be about to leave office, Buchan-Hepburn called on the prime minister on the evening of June 23. This time, there was no attempt to cushion Churchill's feelings. There was no pretense of approaching him out of concern for his health, no willingness to indulge his wish to remain in office until after the coronation, and no sop to his dignity such as the proposal that he transfer to the Lords had been. This was strictly about Churchill's performance and the party's ability to win elections. The chief whip conveyed the ministers' opinion that he ought to retire at once or at least set a firm date for his resignation. Churchill refused to do either. On this and similar occasions, he affected a frosty exterior, but, as his wife and children understood, he was desperately hurt by the efforts of fellow Conservatives to force him out.

Later that week, Churchill took revenge on Eden, though the latter had been careful to play no visible role in the machinations on his behalf. Eden's American and French counterparts, Dean Acheson and Foreign Minister Robert Schuman, were then in London. Before their meetings concluded, Churchill invited the trio to lunch. In the course of the afternoon, Churchill steered Acheson to the window of the sitting room where they had all gone to have a glass of sherry. Signaling by his tone of voice and by several conspiratorial pokes that he was about to have some fun at Eden's expense, Churchill announced that he planned to tear out a row of poplars along the garden wall because they spoiled the view. His remarks disturbed Eden, who broke off his talk with Schuman elsewhere in the room. Eden informed Churchill that he could not possibly remove the trees.

"Why not?" said Churchill. "I live here, don't I?"

Eden solemnly reminded the prime minister that occupying Number Ten was not the same as owning it.

"Ah!" said Churchill, coming in for the kill. "I see what you mean. I'm only the life tenant. You're the remainderman."

The intriguers met again to plan their next move. Their deliberations became pointless and even a bit absurd when Eden suddenly fell ill at a meeting with Acheson on June 27. He developed a fever and was diagnosed with jaundice. The Foreign Office announced that he was likely to be out for a month. Days after the Edenites had asked Churchill to step aside, he lustily threw himself forward in the House of Commons to take Eden's place in the debate on Korea and the Far East. On July 1, Churchill, attired in a short coat and pale gray trousers, gave the government's reply to a critical motion tabled by the opposition. Churchill "fairly pulverized his attackers," Chips Channon recorded, "rarely has he been more devastating." There was a sense at the time that the prime minister's lively riposte was aimed not just at Labour but also at those in his own party who thought him too old to carry on.

Churchill was not always so successful, and eventually the strain of these weeks began to show. During his final address in the House of Commons prior to the summer recess, he was unable to maintain his composure when young Labour members heckled him as he struggled to survey the dire economic situation. By the time Churchill sat down, he was flushed and furious. Attlee provoked him further by commenting that while Churchill had undoubtedly accomplished great things in his day, the first nine months of his peacetime government could only be described as squalid. When Churchill returned to his room in the House, his appearance so alarmed Colville that he sent for the doctor.

Not long afterward, Churchill was resting at Chartwell when Eden brought some surprising news. Since January, the heir apparent (whose first marriage had ended two years before) had been secretly engaged to Churchill's niece. Clarissa Churchill, the willowy blonde daughter of the prime minister's late brother and sister-in-law, was twenty-three years Eden's junior. They had first met in 1936 when she was only sixteen. At the time, she had been part of the set of Salisbury's handsome teenaged son Robert, who in those days was rather in love with her himself. When Eden and Churchill's niece met again, in 1947, he became her "latest admirer" and she was soon confiding to another admirer, Duff Cooper, that Eden never stopped trying to make love to her. Persistence paid off and Eden finally married into the Churchill family on August 14, 1952. The Churchills gave the happy couple their blessing, hosted a reception at Number Ten (which to some eyes gave the proceedings a "dynastic" air) and presented the newlyweds with a substantial check. But hardly had they gone off on their honeymoon when Uncle Winston made it clear that he had no intention of treating "my Anthony" any better than he had in the past.

Churchill contacted Truman to say that he was taking charge of Foreign Office correspondence in Eden's absence. He went on to propose that he and Truman make a joint approach to the prime minister of Iran in the hope of settling a dispute about Iranian oil that Eden had been working on. In one of a series of messages, Churchill said, "I thought it might be good if we had a gallop together such as I often had with FDR." This was no small suggestion, for if Truman signed a common telegram it would be the first occasion since 1945 that the Americans agreed to take joint action with Britain against a third party.

During Churchill's recent visit to Washington, Truman had resisted efforts to replicate the Roosevelt-Churchill relationship. Now, as he had done in advance of Potsdam, the president insisted that he wanted to avoid the impression of ganging up. Churchill replied, "I do not myself see why two good men asking only what is right and just should not gang up against a third who is doing wrong. In fact I thought and think that is the way things ought to be done." This time at least, Truman took his point and consented to speak in a single voice with Britain's leader. Colville noted in his diary that Eden, completing his honeymoon in Lisbon, was furious when he learned of Churchill's coup. "It is not the substance but the method which displeases him; the stealing by Winston of his personal thunder."

Churchill, meanwhile, was in a state of euphoria. The significance of the joint action went far beyond Iran. He saw it as a test run for a unified Anglo-American approach to Stalin after the election. Truman had already announced that he would not be a candidate for re-election, and at this point Churchill calculated that an Eisenhower presidency was probably his best hope of facing Stalin again with American power on his side. Whatever differences Churchill may have had with him in the past, Eisenhower was an internationalist. That made for a stark contrast to the isolationism of his chief rival for the Republican nomination, Senator Robert Taft of Ohio. When Eisenhower embarked on the pursuit of the presidency, he had the image of, in Sir Oliver Franks's words, "the best man to win the peace."

But something happened to Eisenhower on the path to power. After he beat out Taft, he made significant compromises in the name of unifying the Republican Party. He veered sharply to the right, he refused to denounce McCarthy, and his foreign policy pronouncements took on a new bellicose tone. He spoke—some thought recklessly—of US action to "liberate" Soviet satellites. Truman accused Eisenhower and his chief foreign policy adviser, John Foster Dulles, of playing "cheap politics" and of increasing the risk of war in a cynical effort to gain votes. The previous summer, when Eisenhower took the lead over Taft, Churchill had expressed relief to his wife. By the time Eisenhower won a landslide victory over Adlai Stevenson in November 1952, Churchill had come to view him very differently. "For your private ear, I am greatly disturbed," Churchill told Colville of the outcome of the presidential election. "I think this makes war much more probable."

In any case, Churchill had been waiting a long time for the American election to be over, and as far as he was concerned there was not a moment to be lost. It was not solely his age and health that worried him. It was also Stalin's. He expected that the international situation would become much more perilous when Stalin's sycophants began to scramble for the succession after he died.

Immediately, Churchill dictated a message to Eisenhower: "I send you my sincere and heartfelt congratulations on your election. I look forward to a renewal of our comradeship and of our work together for the same causes of peace and freedom as in the past." Churchill transmitted the text to the Foreign Office with instructions that it be dispatched at once, as well as made public. Eden, when he read the message, was quick to object. Eisenhower had had some vile things to say about Truman during the campaign, and Eden worried that Churchill's effusive tone might offend the outgoing president. Besides, Eden was far from pleased to see

Churchill try to renew anything with Eisenhower. That sounded dangerously more like a beginning than an end. The Foreign Office advised that Franks be consulted about the wording. The recommendation made Churchill very cross and he insisted on sending the telegram precisely as he had written it.

Eden was rather cross himself afterward when he went off to a luncheon at the Austrian embassy. In "a sudden burst of nerves and temper," he confided to a fellow guest that he and Churchill were on bad terms. "I get all the knocks," Eden complained to Chips Channon, whom he had known at Oxford. "I don't think I can stand it much longer."

Eden's nervous agitation persisted when, two days later, he flew to New York as head of the British delegation to the United Nations General Assembly. En route, despite being heavily drugged, he suffered an attack of severe abdominal pain. He could not sleep and had to be injected by his detective. While he was at the UN, he conducted meetings with both Dulles, who sought to allay British concerns about the foreign policy goals of the new administration, and Eisenhower. To Eden's surprise and delight, Eisenhower did not even mention Churchill. Instead, he talked entirely of working with Eden, who emerged from the meeting convinced that Eisenhower saw Churchill as already essentially out of the picture. Afterward, an emboldened Eden spoke to his private secretary of the Cabinet appointments he intended to make as soon as he became prime minister. On the flight home he continued to speak of all that he planned to do when he took over from Churchill.

Back in Britain, Eden learned that Churchill had been speaking to Clarissa Eden of his retirement. Churchill had suggested to his niece that he wanted to give up and that he was simply looking for the right opportunity. Mrs. Eden asked her husband to be gentle and allow Churchill to travel to Washington before he left office. The question of one last trip came up again presently when he and Churchill conferred. It seemed to Eden that Churchill actually "begged" to be allowed to go. At other times, Churchill appeared almost to wink when he talked of handing over. Shortly after Eden's homecoming, he and Clarissa were included in an intimate family dinner in honor of Uncle Winston's seventy-eighth birthday. The other guests were Churchill's daughters Mary and Diana, his son Randolph, and their spouses. A great many gifts had poured in, and the prime minister reported that he had even received some "rejuvenating" pills from the US. He said he dared not try them: "Think how unfair it would be to Anthony."

The question of when precisely Churchill planned to leave office was one Salisbury insisted must be answered without delay. Salisbury was not pleased when he learned about what was being billed as the prime minister's farewell trip to America. Many times in the past, Churchill had made similar vague assurances. Yet here he was, still in place, still in power. The day after Churchill's birthday, Salisbury had observed him at a reception for Commonwealth prime ministers, where he seemed sadly unable to follow the business of the conference. One participant described Churchill's lack of comprehension as "pathetic." Was it not his colleagues' duty to ensure that he relinquish his post? Eden was due to visit Churchill at Chequers, and Salisbury advised him to be blunt. He said that if Eden failed to extract a specific date he would convene a ministers' meeting (exclusive of the heir apparent) to discuss a new plan of attack.

In advance of the confrontation, Eden's private secretary sounded out Lascelles about current thinking at the palace. Shuckburgh let it be known that both the president-elect and Secretary of State-designate Dulles had communicated a distinct preference for working with Eden. But when Eden's man subtly questioned the usefulness of another visit to the US by Churchill, he discovered that the prime minister had already been in to enlist the Queen's support.

At last, on Sunday, December 7, Eden told Churchill that he must know something of his plans. As he had done many times before, Churchill solemnly suggested that it was his intention to hand over to Eden in the fullness of time. When exactly, Eden pressed, would that be? Both men remained silent for about a minute before Churchill remarked on there being things he could say, speeches he could make, so much more easily, were he no longer prime minister. This, too, sidestepped Eden's question, and there followed another long, tense silence. (In conversation, Churchill had a way of going dead silent at the end of a sentence, then recommencing after a few seconds. The effect could be unnerving. Ought one to speak into the silence, or ought one to wait?) Finally, it was Eden who blinked. He changed the subject once again, and the question of the succession was dropped.

When Churchill contacted Eisenhower to say that he intended to take a holiday to Jamaica and to suggest that they arrange to meet prior to the inauguration, Cabinet ministers fretted that he had revealed himself to be an "old man in a hurry" and that his premature request might irritate the president-elect. Their concerns soon appeared to be unfounded. Eisenhower replied immediately, and he sounded almost as eager to see Churchill as the prime minister was to see him: "I am delighted at the thought of a meeting with you in the first half of January." Eisenhower, naturally, could not come to Churchill in Jamaica, but he suggested they arrange to meet in New York at the home of Bernard Baruch. "This would not be as satisfactory as a longer and more leisurely visit over our respective easels," said Eisenhower, who shared Churchill's love of painting, "yet it would be, from my viewpoint, far better than no meeting at all."

Eisenhower pressed all the right buttons. When he noted that he had managed to avoid formal meetings with representatives of a number of world governments, he was flattering Churchill's conviction that Britain had earned the right to be treated differently from other countries. And when he asserted that the future of the free world required close understanding and cooperation between the English-speaking nations, he was giving a reassuring nod to the concept of the unity of the English-speaking peoples that was so dear to Churchill's heart.

Churchill arrived in New York on January 5, 1953, full of hope that after eight years of struggle he was finally, assuredly, on his way back to the summit. To the distress of certain intensely interested parties at home, he announced that he did not plan to retire until he was "a great deal worse and the Empire a great deal better" and that he was confident his best work was still ahead.

What Churchill did not yet know was that in spite of his warm messages, Eisenhower was determined not to allow him to recreate the close partnership with the White House that he had enjoyed during the war. In his diary, Eisenhower called the prime minister's hope of establishing such a relationship "completely fatuous" and he again expressed the wish that Churchill "would turn over leadership of the British Conservative Party to younger men."

Eisenhower's feelings about Churchill had not been softened by the role that the Berlin controversy had played during his campaigns for both the Republican nomination and the presidency. Eisenhower had been dogged by questions—from the right wing of his party, from potential supporters and contributors, from Republican delegates, and from the voters themselves—about his 1945 decision to let the Russians reach Berlin first. In the opening salvo of his effort to secure the Republican nomination over Taft, he told an audience in Detroit that one of the questions most frequently asked of him was, "General, why didn't you take Berlin?" Again and again, in letters, private conversations, and speeches, he had made the case that "no political directive" had ever been given to him to take Berlin, that it had been a "destroyed city," and that he had had no role in the unsound political decisions that had already set the lines of occupation in the Soviets' favor. The first and second points, while true in themselves, were not quite the whole story. Churchill had implored Eisenhower to take Berlin and had written to impress upon him the political and symbolic importance of capturing it, Stalin's lies to the contrary. The third point was Eisenhower's old trick of shifting the blame for postwar problems to the British prime minister. At times in his campaign statements, Eisenhower seemed almost to be trying to convince himself as much as anyone else that he had not simply been duped by Stalin. He could get away with making these and related claims so long as Churchill's 1945 telegrams to him were not in public circulation. That would change as soon as Churchill published volume six of his war memoirs, it being his custom to print the relevant correspondence verbatim.

Churchill had come to New York to seek the benefit of American power, without which he could not hope to negotiate effectively with Stalin. The initial signs seemed promising when he met Eisenhower in Bernard Baruch's living room, where a painted portrait of Churchill in military uniform hung above the fireplace. Eisenhower spoke with "much vigor" of his intention to parley with Stalin after the inauguration. It seemed like everything Churchill could have wished for—but then suddenly Eisenhower struck his blow. He planned to meet with Stalin alone, cutting Churchill and the British out.

In 1945, Truman had similarly suggested that he wanted to see Stalin by himself. Eisenhower's proposal was considerably worse. Truman at least had indicated that Churchill could join them later. Eisenhower had no intention of allowing the prime minister to participate at any time. He had completely outmaneuvered Churchill, who had often suggested that his reason for clinging to power was the hope that he might yet have a chance to win the postwar peace. Whether or not Eisenhower really meant to see Stalin—during the presidential campaign he had said that he did not think a meeting at this time would help solve world problems—he had effectively blocked Churchill from putting pressure on the new administration by calling for a summit conference. The prime minister was unlikely to demand a meeting from which he knew he would be excluded. If and when Eisenhower announced his intention to talk to Stalin, Churchill, a staunch advocate of top-level talks, could hardly argue publicly that the president ought not to go.

As if to rub salt into the wound by reminding Churchill that he was powerless without the Americans, Eisenhower politely informed him that he was "quite welcome" to arrange to see Stalin separately, if he saw fit, at any time. Eisenhower for his part claimed to be contemplating an announcement in his inaugural address of his willingness to meet Stalin one-on-one in Stockholm. At this, Churchill frantically shifted gears. Suddenly, the "old man in a hurry" was doing whatever he could to slow things down in the hope that he might yet find a way to insinuate himself into the process. He suggested that it would be wiser to keep to generalities in the inaugural speech. He reminded Eisenhower that there was no battle going on and that as a general he could wait for full reconnaissance reports. He pointed out that Eisenhower had four years of "certain power" ahead of him, and he asked whether it might not be a mistake to seem "in too great a hurry."

XV

IF NOTHING CAN BE ARRANGED

Jamaica, January 1953

After stopping in Washington for a farewell visit to Truman, Churchill went on to Jamaica for a two-week holiday. It was not a happy interlude. He devoted most of his time to painting, but he also corrected the last three chapters of the final volume of his war memoirs. The need to revisit the frustrations of 1945 did not help his mood in the wake of New York. In conversation with Colville and other companions, he complained bitterly about Eisenhower, calling him "a real man of limited stature." To the last minute, Churchill did not know whether Eisenhower would go through with his plan to immediately propose a one-on-one meeting with Stalin. In New York, he had seemed to listen when Churchill urged him to hold off, but what he would actually do on January 20 remained in suspense.

Churchill was still in Jamaica when Eisenhower was inaugurated. The inaugural address made no mention of a desire to parley with Stalin, but of course the omission did not rule out some sort of public statement in the near future. It did not bode well that Eisenhower had rejected Churchill's suggestion that rather than return directly to London after his holiday, he come to the White House for a week of leisurely talks. To make matters worse, Eisenhower had not even bothered to turn him down personally. That Eisenhower had chosen to convey the unwelcome message via Dulles further suggested that he meant to keep Churchill and the British at a distance.

Churchill returned to London on January 29 to face growing unrest in his own government. While Churchill was abroad, Eden had assented to a proposal by Harry Crookshank that party leaders stage a new showdown with the prime minister, whose fading powers made him "an increasing liability" to the party. Even as Eden protested that he could play no direct role in an effort to force Churchill to step aside after the coronation, he agreed unequivocally that the effort needed to be undertaken. Eden had lately been subjected to a series of blistering attacks in the Beaverbrook press that he and his supporters suspected Churchill's surrogates (i.e., his son or sons-in-law) or even the prime minister himself of having instigated in the belief that tearing down the likely successor might prolong Churchill's tenure in office. The more Eden was criticized in print, the more Butler seemed to come in for praise. Lascelles assured Eden's private secretary that if Churchill were to die at this point, it was still certain that the Queen would send for Eden. He predicted, however, that that might change by the end of the year. According to Lascelles, were present trends to continue, at least fifty per cent of opinion in the party might favor Butler by then. Frantic that Butler might yet overtake him in the race to succeed Churchill, Eden moved to line up key support. He tempted Macmillan with the post of foreign secretary in an Eden government—if, that is, Macmillan backed him over Butler.

When the retiring US ambassador to Great Britain, Walter Gifford, lunched with Eisenhower on February 12, he depicted Churchill as a powerless old man, whose own inner circle longed for him to hand over. The picture confirmed Eisenhower's impressions in New York, where his strategic move to block the prime minister's efforts to get to Stalin had left Churchill sputtering. The president's reaction to Gifford's report was clear from the entry he made in his diary afterward: it was time for Churchill to retire.

Churchill had no intention of stepping aside, whether to please Eden or Eisenhower, but on his return to London he was still miserably in the corner Eisenhower had backed him into in New York. Soon his situation got worse.

During the night of March 1–2, Stalin suffered the severe stroke his physician had warned of if he refused to retire. For some twelve hours, the lieutenants whom Stalin had taunted and tormented until the end put off calling a doctor for fear they would be accused of trying to grab power. When at last a decision was taken to summon help, the choices were limited. Vinogradov was being tortured in an effort to make him confess that he was a longtime agent of British intelligence. Other eminent doctors had been jailed as participants in the "doctors' plot" to murder Stalin and other Soviet leaders. The plot had been a figment of Stalin's paranoia but the Soviet people, who had first read about "killer doctors" in *Pravda* two months previously, did not know that. To allay public fears, Stalin's lieutenants issued a communiqué stating that his treatment would be closely supervised by the Central Committee of the Communist Party and the Soviet government. The doctors who were brought in to save the dying Stalin had never treated him before and the pressure on them was excruciating. Though the tyrant lay helpless, his face "contorted," the medical men trembled as they struggled ineptly to examine him.

When Stalin died on March 5, his position did not pass to any individual. Instead there was a joint leadership comprising Chairman of the Council of Ministers Georgi Malenkov, Minister of State for Security Lavrenti Beria, Foreign Minister Vyacheslav Molotov, First Secretary of the Communist Party Nikita Khrushchev, and Minister of Defense Nikolai Bulganin. Ostensibly, the paunchy, unkempt fifty-one-year-old Malenkov occupied the first place and Beria the second. Still, Malenkov's colleagues—especially Molotov, whom Dulles compared to Machiavelli, and whom a good many observers in the West had been betting on to come out on top—failed to accord him the degree of authority they had unhesitatingly accorded to Stalin. Already, the others were watching Malenkov "like hawks" for any indication that he was not up to the job.

What did the change mean for the last surviving member of the original Big Three? There was reason to think that the death of Stalin might be the final nail in Churchill's political coffin. Churchill had long conceived of a parley at the summit as a personal rematch with Stalin. He had maintained, and seemed sincerely to believe, that his hard-earned visceral sense of the tyrant's cast of mind would smooth the way to agreement and that that alone was ample justification for keeping his would-be successors waiting. Nevertheless, no sooner had Stalin been laid out beside Lenin than Churchill used his demise as an excuse to try again with Eisenhower. Malenkov became Churchill's new Moby-Dick.

On March 11, the night after Stalin's funeral, Churchill drafted a telegram to Eisenhower. He began by saying that he was sure everyone would want to know whether Eisenhower still contemplated a meeting with the Soviets. "I remember our talk at Bernie's when you told me I was welcome to meet Stalin if I thought fit and that you understood this as meaning that you did not want us to go together, but now there is no more Stalin I wonder whether this makes any difference to your view about separate approaches to the new regime or whether there is a possibility of collective action." When Churchill composed this message, he was acting on the assumption that Eisenhower had not wavered from his intention, stated emphatically in New York, to sit down with the Soviets. But when Eisenhower's reply came in the following day, Churchill rethought his whole approach. Not only did Eisenhower refuse to act jointly with the British, he also questioned the wisdom of top-level talks in general. In New York, Eisenhower had portrayed himself as eager to go to the summit; now, he seemed anxious to avoid a meeting with the Soviets on the grounds that it would merely give them another opportunity to take advantage. Suddenly, it was evident that the man who had allowed himself to be duped by Stalin in 1945 was afraid of being made a fool of again. Once Churchill grasped that Eisenhower had been bluffing in New York, he was no longer precluded from moving actively and publicly to arrange a top-level conference.

Determined to force Eisenhower to the summit, Churchill, without telling the president, set to work building momentum toward a leaders' meeting. He began by seeking to throw Eden together with his counterpart in the new Soviet government, Molotov, who was also the one member of the new team with whom both Churchill and Eden had had extensive prior contact. In discussions with British colleagues, Churchill made no effort to conceal his view of an Eden-Molotov encounter in Vienna as only an "interim objective" or of his hope that it would lead to a top-level meeting with Malenkov, attended by both Eisenhower and himself. Under different circumstances, Eden would likely have balked at the prime minister's proposal that Eden take the lead in a new Soviet initiative. Encouraging Churchill's summit dreams and giving him a reason not to quit were the last things Eden wanted to do. Nevertheless, a chance to claim the limelight at a particularly difficult point in his career seems to have proven irresistible to Eden, who was "very keen" from the first.

The indications from the Kremlin, meanwhile, suggested a desire to reduce East-West tensions, though to what end no one could be sure. On March 15, Malenkov made a speech in which he maintained that there was no problem between the two sides that could not be resolved diplomatically. A Soviet offer to comply with previous requests from London that British civilian prisoners in Korea be released was soon followed by larger gestures, such as the acceptance by the Korean Communists of a longstanding UN proposal for the reciprocal release of sick and wounded prisoners and a call for the resumption of the suspended armistice talks at Panmunjom. From the time Churchill moved back to Number Ten in 1951, the Korean War had been a major obstacle to US participation in a summit meeting. Whether or not that obstacle was finally about to come down, Churchill was eager to use the Communist offer on Korea as an excuse to propose the Eden meeting to Molotov.

It quickly became apparent, however, that Eden's health ruled out a meeting any time soon. On April 4, the foreign secretary learned that he required an immediate operation on his gall bladder. The surgery would put him out of play for three to six weeks. Initially, the delay did not seem to worry Churchill. The astonishing news from Moscow that Vinogradov and other jailed doctors had been released and that the charges against them had been fabricated and their confessions obtained improperly (a euphemism for torture) suggested to Churchill that time for the moment was on his side. It struck him that in a dictatorship, an admission that the government had lied could be the beginning of the end. On the theory that the most dangerous moment for evil governments is when they begin to reform, Churchill calculated that he could sit back and let things develop until Eden recovered.

On April 6, however, he learned that Eisenhower had been quietly planning a move of his own. The president sent word that he was contemplating a major foreign policy speech. There had been considerable upset in Washington with the impression Malenkov's speech had made in the world. Eisenhower, in particular, had been distressed that he had missed the opportunity to take the lead with a "big" speech of his own. His forthcoming address on world peace was designed to seize the initiative from Moscow. Eisenhower also assumed it would serve as the defining statement of the West's post-Stalin Soviet policy.

Three days later, a draft of the president's speech was delivered to Churchill. When he read it he was appalled. As he saw it, Eisenhower's speech, scheduled for April 16, threatened to halt the flow of hopeful gestures from the East. Deeply skeptical of Soviet intentions, Eisenhower planned to argue that rhetoric was one thing, concrete acts another. The speech enumerated the specific steps Eisenhower expected the Soviet Union's new leaders to take in Korea, Austria, Germany, and elsewhere in the world to prove their sincerity before he could even consider meeting them.

For Churchill, the most important action to take at this point was to make direct contact with Stalin's successors and to start talking informally and unconditionally. In his mind, what was needed was an "easement" of tensions, a period when one small agreement might lead to another, rather than a single all-encompassing agreement. Churchill hoped that in Stalin's absence the Soviet Union would allow itself to become a little less isolated and therefore more pervious to Western influences, and he worried that Eisenhower's demands might send the new leaders back into their shell just when they had been showing signs of a desire to emerge. Not for the first time, Churchill deplored what he saw as Eisenhower's fatal lack of strategic thinking.

He reacted immediately. On the same day he read Eisenhower's draft address, he offered to let the president see a draft of something he had written. Churchill, who had previously announced that he would respond to *Crusade in Europe* in the relevant volumes of his own memoirs, sent word that the final volume was to appear later in the year. He explained that the book spanned the period from the launching of Overlord to the Potsdam Conference, a time of almost unbroken Allied successes "but darkened by forebodings about the political future of Europe which have since been shown to have been only too well founded." In the guise of asking whether Eisenhower wished to vet the numerous references to him in the manuscript, Churchill implicitly reminded the president of how wrong he had been about Berlin.

Two days later, Churchill followed up with a telegram beseeching Eisenhower to postpone his speech. He argued that the apparent change of mood in Moscow was so new and so indefinite and its causes so obscure that there was not much risk in allowing things to develop. "We do not know what these men mean. We do not want to deter them from saying what they mean." Citing the Kremlin's about-face with regard to the doctors' plot, Churchill said he did not want the world to think Eisenhower's speech had interrupted a "natural flow of events" that might themselves lead to the undoing of Soviet Communism.

When Churchill's second message arrived in Washington, Eisenhower seemed to waver. Clearly, the controversies of 1945 continued to eat at him. Twice since Stalin's death, Eisenhower had erupted weirdly at National Security Council meetings, claiming that had Stalin been able to act freely at the close of the war, the

Soviet Union would have gone on to establish more peaceful and normal relations with the rest of the world. As if he were trying to convince himself that he had not been wrong to swallow Stalin's assurances, he insisted that the Soviet leader had never enjoyed complete power and that he had had to come to terms with other members of his ruling circle. Still, in conversation with speechwriter Emmet John Hughes and other advisers on April 11, Eisenhower was palpably nervous about making the same mistake twice by failing to listen to Churchill. He said, "Well, maybe Churchill's right, and we can whip up some other text for the occasion."

Hughes and certain of the others appealed to Eisenhower's ego by arguing that Churchill's motive for requesting a postponement was "to guard and reserve for himself the initiative in any dramatic new approach to the Soviet leaders." Determined never to have to play second fiddle to Churchill, Eisenhower finally sent word that the speech could not be put off. If Churchill wished to propose any specific changes, he ought to cable his list at once. Eisenhower stipulated that while he could not agree in advance to be guided by all of Churchill's suggestions, he would certainly consider them "prayerfully."

When Churchill received this message, Eden was at the London Clinic waiting to undergo a cholecystectomy. On the eve of the operation, the two men worked frenziedly to try to minimize the damage Eisenhower was about to do. They urged the president to combine his re-assertion of inflexible resolve with some balancing expression of hope that the world had entered a new era. Though Churchill had previously seen no problem in Eden being out of commission for several weeks, suddenly it seemed crucial that he be well enough to face Molotov at the earliest possible date lest the momentum toward a summit meeting be lost. Churchill lectured Eden's surgeon, Basil Hume, on the grave responsibility that had been placed in his hands and warned that nothing must be allowed to go wrong. He demanded constant updates and concerned himself with the most infinitesimal medical details. At times, it seemed almost as if the prime minister was about to insist on donning a surgical mask and cutting out Eden's gall bladder himself.

Churchill's efforts at micromanagement had the effect of completely unhinging Hume. On the day of the operation, Eden was under anesthesia when a further communication reached the operating theater to reiterate the prime minister's interest and anxiety. By this time, Hume was so flustered that the surgery had to be postponed for almost an hour. At length, Eden spent over three hours on the operating table and what ought to have been a routine procedure became a fiasco. The knife "slipped," the bilary duct was inadvertently cut, and Eden lost a bucket of blood and nearly died. He faced a second operation to repair the damage of the first.

Though Eisenhower had agreed to tone down his speech somewhat, the final text persisted in demanding positive evidence of the Kremlin's desire for peace before any talks could take place. Hardly had Eisenhower delivered his speech on April 16 when Churchill undertook to spin the president's remarks. Speaking in Glasgow the next day, he portrayed Eisenhower's address as a "massive and magnificent" statement of the West's peace aims and downplayed, if not altogether ignored, its emphasis on tests and conditions. He stressed that Eisenhower had "closed no door" on sincere efforts to achieve world peace. When Dulles, speaking in Washington, painted Eisenhower's comments as a challenge to the Soviets, Churchill was quick to cast doubt on the secretary of state's interpretation. Churchill insisted in the House of Commons on April 20 that he did not read Eisenhower's speech as a challenge and that he did not expect Moscow to give "an immediate categorical reply to the many grave and true points which his remarkable and inspiring declaration contained." After the prime minister slathered on additional praise for Eisenhower, he expressed the hope that there might soon be a top-level meeting of the principal powers involved in the Cold War. Eisenhower certainly had not said anything about this, though in Churchill's telling it was almost as if he had. Churchill then sent off a transcript of his remarks in the Commons to the British embassy in Moscow to be passed on to Molotov. This was how he wanted the Kremlin to perceive Eisenhower's words. These were the positive lines along which he wanted the Soviets to be thinking.

On April 21, he offered Eisenhower a chance to go along with the spin. He sent on a transcript of his remarks in the Commons, along with a message asking what the president thought should be the next step. Churchill was not bashful about giving his view: "In my opinion the best would be that the three victorious Powers, who separated in Potsdam in 1945, should come together again." He recalled that in New York, the president had spoken of meeting Stalin in Stockholm and he suggested that that might be just the place for the three leaders to meet now.

"If nothing can be arranged," Churchill continued, "I shall have to consider seriously a personal contact. You told me in New York that you would have no objection to this."

In New York, it had been tacitly understood that seeing Stalin separately (without the backing of American might) was precisely what Churchill did not want to do. Three months later, when it was Churchill who raised the possibility of going solo, the suggestion had the odor of a threat.

This shift in tone persisted the next day in the House of Commons, when Churchill followed up his private suggestion to Eisenhower that Britain might be about to take an independent line. The prime minister had lately been under considerable pressure from the Labour front bench to make good on his campaign pledge to arrange a meeting with the Soviets. In reply to Opposition members who heckled him for being overly complimentary to Eisenhower and who asked for assurance that Churchill did not plan to leave the initiative entirely in American hands, he asserted his and his country's independence: "I do not think, looking back over a long period of peace and war, I have ever, so far as I had anything to say in matters, been willing to accept complete initiative from the United States."

That spring, it was widely noticed that the seventy-eight-year-old Churchill seemed suddenly in better fettle. Lord Swinton, the secretary of state for Commonwealth relations, toasted Churchill's renewed vitality. Violet Bonham Carter thought he appeared "extraordinarily well and buoyant—a different creature from what he was last year." Though he was doing Eden's work in addition to his own, he seemed "not the least overburdened." On the contrary, the arrangement seemed to exhilarate him. Churchill savored the irony that he seemed only to get better while everyone around him was going down: Salisbury had been sick, and Macmillan had recently spent a week in hospital with gall bladder troubles of his own. Eden would probably be away from his desk for several months, if not more.

Protesting that he could not bear a sick man, Churchill refused Eden his request to stay involved from home. Previously, Moran had been at pains to lessen the prime minister's load. Now, Churchill himself had gleefully added to it. Worried that, appearances to the contrary, Churchill was heading for a fall, Moran warned on April 24 that while Churchill could probably manage both jobs for a brief period, burdening himself indefinitely was surely unwise. Churchill laughed off his concerns and assured him it was not as much work as he thought.

At Windsor Castle that evening, the prime minister knelt at the feet of the young Queen to be made a Knight of the Garter. In the aftermath of the Labour landslide in 1945 he had rejected that high honor because, "I could not accept the order of the Garter from my Sovereign when I had received the order of the boot from his people." Eight years later, Churchill rejoiced that he was again at one with his countrymen. When Queen Elizabeth tapped his shoulders with a ceremonial sword and he rose as "Sir Winston," some observers mistook his decision to accept a knighthood as a sign that he intended to retire after the coronation.

The White House certainly would not have been miserable to see him bow out. When Eisenhower's response to Churchill's latest telegram reached London on April 27, it was far from what the prime minister had been hoping to hear. Now, Eisenhower was the one who spoke of not rushing. He altogether rejected the idea of a summit meeting at this time. As far as what he had said previously about Churchill being welcome to make personal contact with Stalin, he noted that the situation had "changed considerably" in the interim. The president no longer thought it appropriate for Churchill to talk to the Soviets on his own.

Eisenhower's stance enraged Churchill. As Churchill saw it, the death of Stalin had presented the West with what was perhaps a unique opportunity to make a fresh start with the Soviets. Eisenhower seemed to be about to squander that opportunity. By Churchill's lights, the president's speech had already provoked a most unfortunate countermove from Moscow. In an apparent response to the American effort to seize the diplomatic initiative, Molotov proposed five-power talks that would also include France and the Chinese People's Republic. On April 28, Churchill complained to the Cabinet that five big powers at the table would be too many. If there was to be a meeting, Churchill wanted it to be limited to the Soviets, the Americans, and the British, "who could take up the discussion at the point at which it had been left at the end of the Potsdam Conference in 1945." Just when things had been going so well, suddenly everything threatened to end in stalemate. Churchill blamed it all on Eisenhower. Exasperated, he told the Cabinet that he wished the president "hadn't started this off by making his speech." To his mind, once again Eisenhower was undermining his ability to deal with the Soviets. Churchill was determined not to let him succeed a second time.

After a follow-up operation on April 29 left Eden even worse than before, Churchill recast his plan to send the foreign secretary to see Molotov as the prelude to a top-level encounter. He now intended to wear both hats and take both meetings—first with Molotov, then with Malenkov—himself. And instead of Vienna, he was ready to volunteer to go to Moscow. On May 4, he sent Eisenhower his draft letter to Molotov and waited for the fireworks from Washington. Churchill included no explanation or justification for his decision to charge ahead in defiance of the White House, only a line to say that he was thinking of going for three or four days in the last week of May. Eisenhower promptly advised him not to go and expressed astonishment at his willingness to meet the Soviets on their home ground, which they were likely to take as a sign of weakness. Churchill retorted that he meant to proceed in any case. This time, he did not merely ignore the president's objections; he disputed them point by point. He argued that the Western allies would gain more immediate good will by going as Moscow's guests than they would lose by seeking to court the Soviets. Taunting Eisenhower, whose refusal to face the Soviets he attributed to fear, Churchill asserted that he was "not afraid" of risking his own reputation if he believed there was a chance of advancing the cause of peace.

Churchill went on to suggest that the only way for Eisenhower to stop him was to join him. "Of course, I would much rather go with you to any place you might appoint and that is, I believe, the best chance of a good result." His sole concession was to defer to Eisenhower's concern that by going to Moscow in May Churchill might jeopardize the president's budget in Congress. Churchill promised to put off his trip until late June, after the coronation. "Perhaps by then," he suggested, "you may feel able to propose some combined action."

Four days later, Churchill turned up the heat on Eisenhower. He opened a two-day foreign affairs debate in the House of Commons on May 11 by making their disagreement public in a powerful speech that recalled his greatest wartime efforts. After declaring that the US would have done well to have heeded British advice in 1945, Churchill made a dramatic announcement: "It is the policy of Her Majesty's Government to avoid by every means in their power doing anything or saying anything which could check any favourable reaction that may be taking place and to welcome every sign of improvement in our relations with Russia."

The previous month, Eisenhower had undertaken to define the West's Soviet strategy. Now, the prime minister boldly laid out his own very different approach in full view of the entire world. In a stunning challenge to Eisenhower, Churchill called for immediate talks at the highest level without agenda or preconditions, and he goaded the Americans by insisting that he could not see why anyone should be "frightened" of a summit meeting. At worst, he said, the participants would have established more intimate contacts. At best, the world might have a generation of peace.

In the days that followed, the world press blazed with news of an open rift between London and Washington, and of Churchill's surprise bid for "the diplomatic leadership of the Western world." In a press conference on May 14, Eisenhower rejected Churchill's proposal by reiterating that he would need to see hard evidence of Soviet intentions before he could consider top-level talks. His anger at the position Churchill had put him in was on display at a meeting of the National Security Council on May 20. Warned by US ambassador to the United Nations Henry Cabot Lodge that unless he did something quickly he would lose the initiative gained in his "great" speech of April 16, Eisenhower shot back that Churchill's maneuvers to get a summit meeting made him wonder whether the prime minister's "faculties and judgment were not deteriorating."

Many Britons, by contrast, saw the May 11 speech as evidence that their leader was again at his best. Having lived through two world wars and having experienced what it means to be attacked at home, they were eager to test every possible chance for peace. Though Churchill's speech had its detractors, especially within his own party, even his critics could not deny that the broader public strongly backed him on this. Alone among Western leaders, Churchill appeared to have boldly seized the moment. As Viscount Stansgate said in the House of Lords, "Sir Winston Churchill made a speech which suddenly put into the hearts of the people of this country that feeling of proud confidence in his leadership that we had at the time of the Battle of Britain." Churchill's declaration of diplomatic independence dovetailed beautifully with the wave of renewed patriotic fervor associated with the coronation. The British were tired of always having to follow America's lead in world affairs. They took pleasure in the spectacle of Churchill taunting mighty Washington and they delighted in his insistence that the US would have been wise to heed British advice at the end of the war. The speech also earned high marks throughout Europe, where popular sentiment favored a summit meeting.

Eisenhower, meanwhile, knew what the world did not—that if he refused to participate in a summit meeting, Churchill planned to see the Soviets alone in June. Should that happen, Eisenhower would find himself cast as an obstacle to peace, rather than its champion. Desperate to regain control of the situation, Eisenhower leaped at a request by French Premier René Mayer on May 20 for an early personal meeting with his American and British counterparts. Given Churchill's predilection for three-power talks on the old Churchill-Roosevelt-Stalin model, the French were nervous that they could find themselves excluded from a summit meeting where the fate of Germany was decided. By arranging to confer with Eisenhower and Churchill beforehand, Mayer sought to ensure his country's place at the table. Eisenhower called Churchill that night at about 11 p.m. London time. He said that Mayer needed their answer by the next day when he hoped to announce the allies' talks to the French National Assembly.

Churchill loved the idea of a meeting, but he pretended not to hear when the president suggested that it take place in Maine. Eager to be perceived as hosting the conference, Churchill suggested that the leaders assemble in Bermuda—on British territory. Though Eisenhower guessed that Churchill had heard him perfectly well, he tentatively accepted Bermuda as the venue. The president did, however, take care to stipulate and confirm in writing that the Bermuda talks were in no way preliminary to four-power talks with the Soviets. He saw the meeting, and he made it clear that he expected Churchill to see it as well, strictly as a forum in which to discuss their common problems. Churchill solemnly agreed.

When Churchill faced the Cabinet the next day he was "in a mood of almost schoolboy enthusiasm." Far from having Churchill in his hand, Eisenhower had presented him with an opportunity both to portray the British as driving events and to make his case for a parley at the summit directly to the president. A number of ministers were not without ambivalence. Salisbury in particular was "frankly skeptical" about Churchill's claims of a change of heart at the Kremlin. Might it merely be that the Soviet Union's new rulers were seeking "a quiet time in which to find their feet"? He was also distressed that Churchill had failed to consult the Cabinet in advance of his summit proposal. Nevertheless, when Churchill recounted his telephone conversations with Eisenhower, Salisbury chimed in approvingly, "Eisenhower has followed the PM's lead."

Salisbury saw Eisenhower's decision to call an allies' meeting as "a direct result" of the May 11 speech. While he persisted in his disapproval at Churchill's failure to check in first, he could not help but be pleased that Britain, and therefore the West, had gained the initiative in foreign affairs. A "completely negative attitude," such as Eisenhower had taken, would have left the initiative with Moscow. That, Salisbury believed, would have been very bad indeed.

News of the conference to be held in Bermuda was met with a tremendous reception in the House of Commons later that day. Attlee asked whether it might, perhaps, be a preliminary to talks with Malenkov. Hours after Churchill had assured Eisenhower that they were in full agreement about the purposes of the conference, he sang a different song in public. "Yes, sir, it is my hope that we may take a definite step toward a meeting of far greater import," Churchill said, to shouts of approbation from nearly every corner of the House. The cheers from the Labour benches appeared to be the loudest despite the socialists' concern that Churchill had stolen their political thunder. With his answer, Churchill transformed the Bermuda conference into precisely what Eisenhower had stipulated it must not be. From this point on, it would be up to the Americans to counter the perception of Bermuda as the prelude to a summit meeting.

Five months had passed since Eisenhower, in New York, had moved to shut Churchill out by declaring that he intended to meet Stalin alone. In the interim, Churchill had argued, cozened, cajoled, threatened, and, finally, muscled his way back into the game.

XVI

THE ABDICATION OF DIOCLETIAN

Buckingham Palace, 1953

Shortly before midnight on June 5, 1953, the prime minister's car pulled up at a side entrance to Buckingham Palace. Three days after the Archbishop of Canterbury had placed the heavy crown of Edward the Confessor upon the head of Elizabeth II, she was to preside over a reception for some two thousand envoys, princes, and other dignitaries, as many of them prepared to leave London. An estimated fifty thousand people waited in front of the palace to see the guests, who wore all manner of uniforms, evening attire, and jewels. Churchill was swanked out in his dress uniform of Lord Warden of the Cinque Ports. He had a cocked hat, a sword, heavy gold epaulets, and gold braid, and his chest glittered with medals and a blinding 150-diamond Garter star.

Inside, as Churchill headed to the white-and-gold state ballroom, the detective who accompanied him suggested that he might want to visit the Gents' first.

"Well," Churchill replied, "just as a precautionary measure you might be right."

When Churchill emerged from the men's room, he came face to face with his longtime political foe Aneurin Bevan, on the way in. Bevan had annoyed not a few people by attending the coronation at Westminster Abbey in a business suit, and he was similarly attired this evening. Churchill, whose fondness for uniforms Bevan had been known to mock, stared at him frostily and said, "You might have taken the trouble to dress correctly on this occasion at least."

Bevan retreated slightly, surveyed him from top to bottom, and called out, "Winston, your flies are undone!"

"Do not concern yourself too much about that," Churchill replied, "dead birds never fly from their nest."

He walked off, doing up his trousers as he went.

Churchill was in a wonderful mood that night, and he was not about to let anyone spoil his pleasure. Nobody truly believed that he would retire now that the coronation festivities were drawing to a close. Instead, he was headed to Bermuda with overwhelming popular support—and then, if all went well, back to the summit.

In recent days, as he played a prominent part in the royal rituals (and as he fought off the waves of fatigue that had nearly prevented him from attending the coronation), he was also working methodically to keep up the post-May 11 momentum. He had leaned on the Soviets to break the deadlock over the prisoner exchange in Korea that threatened to provide the White House with an excuse to reject a summit meeting. He had been in contact not only with Molotov, but also, for the first time, with the little-known Malenkov. He had sold the concept of East-West talks to the visiting Commonwealth prime ministers, so that when he reached Bermuda he could tell Eisenhower that he was speaking not just for himself but also for them.

Even if Churchill had been of a mind to hand over power, he could hardly have done so when the second-in-command whose political interests he had pledged to protect was too ill to replace him. On June 5, Churchill had begun his day with a trip to the airport to bid farewell to an emaciated Anthony Eden as he left for the US to undergo the operation his new American doctor described as the only hope of saving his life. If Eden survived, he was likely to be out of commission for much of 1953. Under the circumstances, none of the Cabinet ministers who had plotted to force Churchill out in favor of Eden would have been pleased to see Churchill go.

That did not mean there was not acute nervousness among them about, in Macmillan's words, the "big game" he was playing. Was it all about to degenerate into a high-risk game of chicken? Churchill was confident that face-to-face he would be able to change Eisenhower's mind. But what if he was wrong?

When they met at Epsom Downs the next day, Macmillan and Salisbury discussed their concerns about what Churchill would do if Eisenhower proved to be as obstinate as he was. Macmillan, who, unlike Salisbury, knew Eisenhower well, was of the opinion that the president "would not trust himself" to go to a high-level meeting. In contrast to Churchill, who preferred to do everything himself, Eisenhower's modus operandi was to work "through" others. That afternoon, the sun shone brightly and by some estimates the crowd was the largest and quite possibly the noisiest in Epsom history. Queen Elizabeth had arrived in the royal train with Prince Philip, Princess Margaret, the Queen Mother, and other members of the royal family. Churchill, who as usual had been late to get started, had had to be driven posthaste on the wrong side of the road to get him to Epsom in time to watch the race from the Queen's box. As it happened, Elizabeth II had a horse of her own in the competition. The co-favorites in a field of twenty-nine were the Queen's chestnut colt Aureole and Sir Victor Sassoon's brown colt Pinza. The latter was to be ridden by the forty-nine-year-old Gordon Richards, whom many racing enthusiasts regarded as the finest jockey of all time. Now approaching retirement, Richards had won more races than any of his peers. One accomplishment had long eluded him, however. He had yet to win a Derby. That changed when Pinza won by four lengths and Richards was called from the winner's enclosure to receive the Queen's congratulations.

In the midst of the excitement, Macmillan and Salisbury talked of Churchill's recent efforts to set the seal on his own career. Salisbury worried that Churchill would go ahead and meet the Soviets even if he failed to win over Eisenhower. In his view, a summit meeting without Eisenhower would be hugely destructive to the Anglo-American alliance. Macmillan opined that that would be "a most serious step" which could not be taken in the absence of Cabinet approval. After the Derby, Salisbury communicated his concerns in a private letter to Churchill. He suggested that the present state of public opinion in the US made a joint approach unlikely. And he argued that even if there was enormous pressure from the British left to go on alone and even if Eisenhower showed "an understanding attitude," Churchill must resist the "great temptation" of a bilateral meeting with Malenkov. "I am sure that the only safe line is to move only on all fours with the Americans, prodding them no doubt at intervals but never advancing without them."

To Churchill's frustration, the Bermuda talks which were to have taken place in the middle of June had to be postponed twice due to the fall of the Mayer government. Finally, though a new regime had yet to materialize in Paris, he and Eisenhower agreed to a firm set of dates. Churchill would arrive in Bermuda on July 6 and welcome Eisenhower there the next day. Presumably by then France would have a new leader, who would be arriving as well.

On Tuesday evening, June 23, a week before he was due to sail, Churchill undertook one last formal engagement, presiding over a large dinner party in honor of Alcide de Gasperi, the Italian prime minister. Though Churchill had seemed fatigued earlier in the day, he held forth with brio, delighting his guests with a witty after-dinner speech about the Roman conquest of Britain. Later, the ebullient host was making his way to the door of the drawing room in what appeared to be good health when he slumped suddenly into the nearest chair. A number of people assumed that he simply had had too much champagne. The art historian Kenneth Clark suspected something else. Clark, who had been asked to the Gasperi dinner because he spoke fluent Italian, instructed his wife to sit beside Churchill while he looked for Mary Soames.

Churchill, barely able to move, clutched Jane Clark's hand and said, "I want the hand of a friend. They put too much on me. Foreign affairs . . ." After that, his voice drifted off.

Christopher Soames dashed to Churchill's side, and moved quickly to bring the evening to an end. He warned Clementine

Churchill that Winston could not walk and that they would have to wait until the room was cleared before they could get him to bed. Signor de Gasperi and the others were told only that their host was overtired. On her husband's instructions, Mary Soames did her best to keep any of the guests from talking to her father, whose speech was slurred and incoherent. At last, Churchill, leaning heavily on Jock Colville's arm, struggled to his room. The doctor who had warned him either to retire immediately or lighten his load if he hoped to avoid another stroke was nowhere to be found. Colville left a message asking Moran to come in the morning.

When Churchill awoke the next day, his mouth sagged on the left side, particularly when he tried to speak, and he remained unsteady on his feet. When Moran examined him, he said that there had been a spasm of a small artery. The episode appeared to be similar to the one in 1949 in the South of France, but it would take a few days to be certain. For the moment all they could do was wait and see how his condition developed.

Immediately, Churchill approached Moran's preliminary diagnosis as he would a negotiation. As he had lately done with Eisenhower, and with countless other opponents through the years, Churchill seized on the words that appealed to him and ignored the rest. Moran had suggested that the incident might be a reprise of 1949; as far as Churchill was concerned, so it must be. The first stroke had been minor. He had had a relatively quick recovery and had gone on from there to recapture the premiership. At a moment when Churchill was finally about to make his case to Eisenhower, he was not prepared to entertain the possibility that this new episode could in fact be the major stroke of which he had long been warned. He was intent on recovering in time to sail on Tuesday.

Hardly had Moran left Number Ten with a promise to come back later when Churchill asked to be helped down to the Cabinet Room. In spite of everything, he meant to preside over a Cabinet. He planned to be seated in his usual chair in front of the fireplace before the others came in so they would not have a chance to see him walk. And he intended to say as little as possible, the droop in the left side of his mouth being more prominent when he spoke. The point was to show himself to his colleagues and to act as if nothing were wrong.

For the most part, he succeeded. Salisbury, who had been at the Gasperi dinner, took his customary place on the prime minister's left and appeared to notice nothing. Butler, who also had been present the night before, registered only that the prime minister seemed more taciturn than usual. Macmillan, though he thought Churchill unusually pale, was not alarmed. Only later did Macmillan suspect the significance of something Churchill said to him in the course of the Cabinet: "Harold, you might draw the blind down a little, will you?" Crookshank alone would later insist that he suspected the prime minister had had a stroke, but if he really thought that, he did not mention it at the time.

Moran returned that afternoon accompanied by Brain. Again, Moran stressed that it would be some days before they could be certain how serious matters were. When Churchill announced plans to answer Prime Minister's Questions in the House of Commons, the doctor warned against it. Obstinacy had gotten Churchill through the Cabinet earlier, but that was no guarantee things might not get worse at any time. Moran cautioned that when Churchill rose to speak, the wrong words might come out—or no words at all. Still Churchill kept insisting he would go. He acted as if to acknowledge the doctor's warnings would also be to accept that this might not be the minor incident he was determined it must be. Moran persisted; and finally, rather than seem to give in, Churchill called for a list of the questions and quickly pronounced them unimportant. He made it clear that he was canceling his appearance not because of anything the doctor had said. He planned to remain in seclusion at Number Ten and wait for the symptoms to pass.

He did not improve, far from it. On Thursday morning, Churchill's speech was thick and his ability to walk had deteriorated. He wanted to attend another Cabinet later in the day, when he was expected to fulfill his promise to Salisbury to fully discuss Bermuda and hear the views of his colleagues. When Moran suggested that it would be best to put this off, Churchill dug in. Only after the doctor mentioned that the drooping left side of his face would attract notice did Churchill relent. Salisbury, Butler, and other ministers had tried to push him out in the past. He was reluctant to let them see his condition and feel emboldened—certainly not when he would soon be better. So he agreed to postpone until the Monday.

At this point, Moran remained unsure of just how the situation would develop. There might indeed still be a chance that Churchill's formidable powers of resilience and recuperation would permit him to sail. But if he were to carry on, it was important that he stay out of sight until he looked and felt better. Concealing his condition was almost impossible as long as he remained at Number Ten. It would all be much easier if he were to hide out at Chartwell. Churchill initially resisted the suggestion. If he were to leave London unexpectedly, would that not create suspicion? Finally he gave in, and Lady Churchill went on ahead to prepare the house for his arrival. To explain the prime minister's sudden departure it was announced that he was going off to prepare for his talks with Eisenhower. The plan was to return for the Monday Cabinet and sail the next day.

At about noon, Churchill emerged from Number Ten in plain sight of any members of the press or public who happened to be outside. Happily, he was able to walk unaided to his car and his condition went undetected. Colville accompanied him on the trip, and en route Churchill gave strict orders that no one was to be told that he was "temporarily incapacitated." He directed Colville to see to it that the government continued to operate as though he were still in complete control. Churchill was going to Chartwell to rest and recover, but by the time they got there his condition had worsened substantially. He could not get out of the car or walk to the house without help. Had he left London just a bit later, people would have seen his incapacity and the game would have been up.

In defiance of Churchill's instructions, Colville contacted three press barons who were the prime minister's friends. Lords Beaverbrook, Bracken, and Camrose came at once to Chartwell, where they hatched a plan to keep the news of the stroke out of the papers. While they were there, the hope remained that if Churchill could be gotten onto the waiting ship he might yet have enough time at sea to recover sufficiently for the Bermuda conference. By Thursday night, it became apparent that in the absence of a miracle there could be no question of his meeting Eisenhower and little, apparently, of his remaining in office. His left side was now partly paralyzed and he had lost the use of his left arm.

When Moran examined his patient on Friday morning, he determined that the thrombosis was spreading. Seventy-two hours before, Churchill had been hoping to move nations and save the world. Now, he found it difficult just to turn over in bed. No amount of willpower or positive thinking had been capable of halting the slow spread of the paralysis. Churchill had been unable to outtalk and outwit his body. The force of facts had reasserted itself: he had sustained a serious stroke and even he acknowledged that Bermuda had to be canceled. Recovery—if he did recover—was going to take much more than a weekend in the country.

Under ordinary circumstances, there would have been no doubt that Eden's moment had come at last. The trouble was that the heir apparent was then in the US recovering from an eight-hour operation that he had been given only a fifty-fifty chance of surviving. Eden was not scheduled to leave his Boston hospital until the following week, when he planned to travel to Newport, Rhode Island, to recuperate at the home of friends. Even if everything went smoothly, it would be at least October before he could return to work. Whether he would ever really be able to resume his burdens was an open question. Eden's American doctor had warned the British ambassador to the US, Sir Roger Makins, that there was only a ten per cent chance of full recovery.

Eden's predicament gave Churchill room to maneuver. Even now, he was determined not to have to resign. Were it to be announced that Churchill had had a major stroke he surely would have to hand over, if not to Eden then to Butler or some other successor. Instead, Churchill decided, in a favorite phrase, to "pig it." He would try to get away with holding on to power by keeping his true medical condition a secret. As he was likely to be out of commission for some time, there would have to be a public explanation, as well as some mechanism for running the government in his absence. To the first end, Moran prepared a medical bulletin that attempted to explain why the prime minister who had fought so hard for a conference in Bermuda would suddenly bow to his doctors' order to cancel it.

Of course, Moran said nothing about a stroke. Instead, he wrote that Churchill had suffered "a disturbance of the cerebral circulation" which had resulted in "attacks of giddiness." Both Moran and Brain signed the statement, which could not be released until after Eisenhower had been notified that Bermuda was off. Thereupon the physicians left Chartwell and the politicians took over.

At times, Churchill could barely make his words understood, yet the wheels in his brain were still spinning furiously. On Friday, urgent messages were delivered to the two adversaries most likely to rush in upon him when they learned he was wounded. Rather than keep them in the dark for as long as possible, he made sure they were among the very first to know. Salisbury was in London preparing to go off to lunch with a friend when he received a note which Colville had written in his own hand the previous evening. Salisbury immediately canceled his engagement and went to Chartwell. Butler was at the bedside of his dying mother when the summons reached him. He too left at once.

The Lord President and the chancellor arrived to find the family in tears. When Churchill appeared for a late lunch, the effects of the stroke were manifest. In halting, often tortured conversation, he frankly laid out the details of his ordeal and raised the question of the future of his premiership. Given the company, Churchill was playing a dangerous game. If at any time either Salisbury or Butler insisted that Churchill's stroke be made known to the country, he would only be saying what was right and proper. If either or both men chose to go public with the information, there was nothing

Churchill could do to stop them. Once the news was out, there would be no way to hang on long enough to recover. He would have little choice but to retire. For the moment, if he hoped to keep his job the only card he had to play was his well-honed understanding of his guests' complicated and conflicting personal interests.

Churchill claimed to want to hold on only until Eden was well enough to return to England. He insisted that when Eden reappeared, probably in October, he would of course step aside. He cast his actions as an effort to serve not his own interests but Eden's. He pretended that, sick as he was, he was determined not to let Anthony down. If Churchill left office while Eden was incapacitated, someone else would get the top job, and that would not be fair to Eden—his "rightful deputy," as he pointedly informed Butler.

But, Churchill went on, he could only protect Eden if he could conceal that he had had a major stroke and if he could devolve his duties to others. As Parliament was due to begin its summer recess at the end of July, there was reason to believe that together he and his lunch guests might bring off the ruse. Churchill asked them to take on the major burden of his duties. He invited Butler to fill in as head of government. In addition to his present responsibilities as chancellor of the exchequer, Butler would preside over the Cabinet and, along with the Lord Privy Seal, Harry Crookshank, share responsibility for making any statements or answering questions in the House of Commons. Churchill asked Salisbury to oversee foreign affairs. Since Salisbury was in the Lords, the foreign secretary's chores in the Commons would fall to Butler as well.

The moment of decision had come. On the face of it, that either Salisbury or Butler would do as Churchill asked seemed improbable. Even cast as a selfless action, the proposal was outrageous. Of course the country ought not to be deceived. Of course the prime minister should step aside now that he was unable to carry out his duties. Salisbury had been scheming to unseat him since 1945. Was this not the perfect moment to force Churchill into retirement? As for Butler, he too had plotted, both with fellow Conservatives and with Labour members interested in a coalition.

Butler had long been frustrated by the wisdom that the Tory succession was fixed. Was he not likely to welcome a chance to seize the leadership at a moment when Eden was out of the country and out of commission? Why should he agree to keep the seat warm for his rival?

Actually, there were some very good reasons. For one thing, Butler's background as an appeaser threatened to make it impossible for him to form a government at this point. While he had his supporters among Conservatives, he also had important detractors, like Salisbury, who had yet to forgive his position in the 1930s. If, however, Butler were to chivalrously step aside, he would build up future credit with the party. Besides, there was always the chance that Eden's health might prevent him from returning to work in the autumn. In that case, Butler, having won the gratitude of the Edenites and having proven his loyalty to the party, might be in a position to step in sooner than expected.

Butler agreed to assist Churchill, but his cooperation would mean nothing if Salisbury refused to go along. Salisbury was a stickler for rules and decorum. It seemed inconceivable that this most punctilious and respectable of men would even consider Churchill's proposal.

Still, the scheme had its attractions. Salisbury had striven for years to install Eden at Number Ten. Now the premiership threatened to fall into Butler's hands. In that case, Eden would hardly be the only one to suffer. Were the prize to go to Rab Butler, Salisbury was unlikely to enjoy the influence he sought in a new Conservative government.

When Salisbury agreed to the cover-up, he was consenting to participate in precisely the sort of behavior for which he had often criticized Churchill. As ever Salisbury preferred to avoid the limelight, so he asked not to be named acting foreign secretary. He would be available to consult on foreign affairs, while minister of state Selwyn Lloyd would assume responsibility for day-to-day Foreign Office business. Churchill had no problem with this arrangement. The important thing was to feel that he had Salisbury safely in the meshes. He also wanted to be in a position to ask one other crucial favor.

Churchill was distraught that as a consequence of his paralysis, the momentum which had derived from his speech of May 11 had been interrupted. It seemed to him, however, that the interruption need not necessarily be fatal. He had yet to tell Eisenhower that Bermuda had to be scratched, but when he did he wanted to be able to say that Salisbury was prepared to travel to Washington to express the British (i.e. Churchill's) point of view on the next step with the Soviets. Churchill was asking Salisbury in effect to represent a policy with which the Lord President privately disagreed. In the guise of protecting Eden's future, Churchill was using Salisbury, really, to safeguard his own. He needed Salisbury, of all men, to help keep his summit proposal alive. He could not hope to secure the last prize without him.

Salisbury signed up to visit Washington in the event that the Americans agreed to a meeting. He did not, however, believe that the medical report, which Churchill planned to send along with his message to Eisenhower, ought to read quite as it did. It was not that he thought Moran had gone too far in his obfuscations; in Salisbury's view, the doctor had not gone far enough. Moran had spoken of a "cerebral disturbance." Salisbury worried that such language pointed all too clearly to a stroke. It was well known that Churchill had been functioning both as prime minister and foreign secretary in Eden's absence. Salisbury and Butler revised the medical report to say simply that Churchill had had no respite from his arduous duties for a long time and was in need of a complete rest; his physicians had therefore advised him to lighten his burdens and rest for at least a month. In the ministers' opinion this was all that the public, or Eisenhower for that matter, needed to be told.

By the end of lunch, Churchill had transformed both adversaries into ardent enablers. This time, rather than set the birds against one another, he had insidiously encouraged them to fly in formation. Despite their clashing purposes, everyone was supposedly doing this for Anthony. Having bought himself time to recover,

Churchill picked up a beaker of brandy with his good arm and gingerly raised it to his twisted mouth.

For all of Churchill's cold tenacity, it soon looked as if his efforts might have been for naught. That evening, his condition worsened to the point that Moran told Colville he did not know if the prime minister would live through the weekend. Churchill continued to decline overnight. On Saturday, he could barely stand when the doctor got him out of bed. Suddenly, his good hand seemed to be affected, provoking concern that his right side was about to stiffen as well. The house was filled with gloom, but somehow Clementine maintained her spirits and concentrated on trying to make her husband and family comfortable. Randolph arrived at mid-day and Sarah was called home from America. Churchill appeared to have given up hope and the end seemed near.

On Sunday he rallied. Though the left arm and leg remained frozen, Moran judged that the thrombosis was no longer on the move. Beaverbrook came to lunch and reminded Churchill of his complete recovery after his previous stroke. He was fibbing, of course, when he claimed that Churchill was hardly as bad as he had been in 1949, but his words seemed to have an effect. After lunch, Churchill refused his wheelchair and, though he required a helper on each side, he insisted on walking to his room. It was the first of many goals he was to set for himself. As he lurched forward, the toes of his lifeless left foot dragged on the carpet. When he finally reached his bed he dropped from the exertion—but he had made it.

At the Cabinet meeting on Monday, Butler disclosed that, contrary to the sanitized medical bulletin released two days previously, Churchill had suffered a serious stroke. When they heard the news, some of the very men who had intrigued to force Churchill out wept openly or fought back tears. Butler warned that the prime minister's condition must be kept secret. He went on to detail the arrangements that had been agreed to on Friday to carry on government business. Salisbury spoke of Churchill's request that he be available to go to Washington. In the interim, Salisbury had heard twice from Dulles, agreeing to an "intermediate" meeting of British, American, and French foreign ministers, which would be "preparatory to a Bermuda meeting later." Salisbury thought he should accept and the ministers concurred.

Cabinet secretary Norman Brook went to see Churchill for himself the next day. On Tuesday, June 30, Brook and Colville dined with the prime minister, who was in a wheelchair. After the meal, Churchill announced that he intended to stand without assistance. The others advised him not to try, but there was no stopping him. When his guests positioned themselves on either side for safety's sake, he warned them away with his stick. Then he lowered both feet to the floor, clutched the arms of his chair and slowly, painfully, purposefully hoisted himself up, as sweat poured down his face. At last, having shown what he was still capable of, he sat and returned to his cigar. Brook viewed the scene as a demonstration of the refusal to accept defeat that had served Britain so well in the war: "As he had done for the nation in 1940, so he did for his own life in 1953. He was determined to recover."

Once Beaverbrook had spoken of complete recovery, Churchill made that his mantra. He could no longer pretend that he had had another minor stroke, but he could focus on all that he had managed to accomplish since the earlier episode. As far as Churchill was concerned, there was no reason it should be any different this time. The previous Friday he had attempted to keep his stroke a secret from Eisenhower. (The medical bulletin had not fooled the President, who guessed immediately that there was something far more serious involved.) Now, Churchill took a different tack. He wrote to Eisenhower on July 1, "Four years ago in 1949, I had another similar attack and was for a good many days unable to sign my name. As I was out of Office I kept this secret and have managed to work through two General Elections and a lot of other business since. I am therefore not without hope of pursuing my theme a little longer but it will be a few weeks before any opinions can be formed. I am glad to say I am already making progress." In the same strain he informed Macmillan, who came to dine the following day, "I have had a stroke. Did you know it was the second? I had one in 1949, and fought two elections after that."

During the weekend of July 4–5, Churchill had a major breakthrough. Exactly one week before, it had seemed as if the old escapologist had finally run out of rope. Now he managed to walk a short distance unaided and he vowed to accomplish a bit more daily. These efforts left him exhausted and in a considerable amount of pain, but he refused to give up, intent on regaining the full use of his limbs and becoming again the man he had been before the stroke. It had long been Churchill's goal to return to the summit; it looked as if he meant to drag himself there if necessary.

As Churchill battled his way back to health, he was also micromanaging the preparations for the Washington talks. He dictated a long, lucid statement that set out what he hoped Salisbury would be able to accomplish on his behalf in the US. Although the European Defense Community proposal had originated in France, the idea of a rearmed Germany in any form remained highly controversial for many Frenchmen. As a consequence, the French had still not ratified the EDC. Churchill was convinced, however, that a rearmed Germany would put the West in a stronger position to negotiate with Moscow. He wanted Salisbury to go to Washington with two key goals: to press the French to ratify the EDC by October, and to get the Americans and the French to agree that as soon as ratification occurred, they would call for a top-level meeting with the Soviets before the end of the year.

On July 6, two days before Salisbury was to fly to the US, he read the prime minister's statement aloud to the Cabinet. The ministers agreed that it would be Salisbury's mission to pursue the aims set out in Churchill's paper. The following night, Salisbury and his wife dined at Chartwell, where Churchill briefed him additionally for two hours. Salisbury by this time had had to agree to a good deal more than he had originally bargained for. The opposition had balked at an arrangement that divided responsibility for foreign affairs, with no one individual clearly in charge. Salisbury therefore had reluctantly accepted the title of acting foreign secretary. He had had no choice if the cover-up was to work.

At this point, it seemed to Churchill that every sign pointed to success in Washington. Eisenhower had written to assure him that he looked upon the cancellation of Bermuda as only "a temporary deferment of our meeting." And now, as the Washington talks were about to begin, the President responded favorably to a reporter's question about whether in view of Churchill's health he might be willing to go to London. At a time when it looked as if Churchill's remaining days in office were limited, Eisenhower could afford to seem gracious. Churchill, however, was quick to pounce on the president's offhand comments. As though a decision had already been made in Washington, he spoke enthusiastically to his doctor of Eisenhower's impending visit and of the favorite dish of Irish stew Clemmie would have prepared for him.

Also propitious from the vantage point of Chartwell was the excellent press Salisbury, a little-known figure in America, received on his arrival. He was billed as a major opponent of the appeasers in the run-up to the Second World War, an image that promised to lend credibility when he made the controversial case for talks with Stalin's successors. There was some perplexity in the US press about Salisbury's rumored differences with Churchill over the usefulness of a parley at the summit. As soon as the acting foreign secretary emerged from the plane at National Airport, he therefore made it clear in conversations with reporters that he had come to argue for early four-power talks.

Before the foreign secretaries had even had a chance to confer, some startling news from Moscow scrambled everyone's calculations. The number two man, Lavrenti Beria, had been arrested as a traitor. His fall raised important questions about the stability of the new joint leadership. With things so unsettled, was this really the moment to talk to the Soviets? Who was to say that Malenkov might not be the next to go? Churchill would liken the Soviet succession struggle to a "bulldog fight under a rug." "An outsider only hears the growling," Churchill declared, "and when he sees the bones fly out from beneath it is obvious who has won." But at this point, nothing was obvious yet, and Salisbury dryly observed that Beria's arrest did not make his task in Washington easier.

At the first session, on July 10, Salisbury moved at once to turn the uncertain situation to Britain's advantage. He nimbly argued that the lack of clarity about events at the Kremlin since Stalin's death was precisely why Churchill had called for early contact with the new Soviet leaders, who were themselves such "an unknown quantity." As Salisbury explained, Churchill saw a meeting as an opportunity to explore their minds in an effort to understand what was really going on in Moscow. Salisbury urged Dulles and French Foreign Minister Georges Bidault at least to agree in principle to the desirability of leaders' talks "when practicable."

Salisbury's counterparts would not hear of it. Where Churchill sought a top-level encounter, the Americans and the French joined forces to propose a meeting at the foreign ministers' level. Where Churchill was adamant on the need to have a card to play with the Soviets in the form of a German army, Dulles and Bidault were ready to see Molotov before the EDC had been ratified. When it became evident that Salisbury was not about to budge them, he cabled London for instructions.

Churchill's blood rose when he learned where things stood. Since 1945, when Truman, Attlee, and Stalin had concluded the Potsdam Conference by handing off the most contentious questions (including Germany) to foreign ministers, Churchill had often made that tactic the target of his shafts; as far as he was concerned, history had amply vindicated him. Yet here were the US and France demanding more of the same. Churchill instructed Butler to convey his view to the Cabinet that any talks were better than none and that Salisbury should ask that the allies at least keep an open mind to the possibility that Churchill, Eisenhower, and Malenkov would meet later.

At London's direction, Salisbury pressed strongly, but to no avail. When in the end the Washington communiqué spoke only of foreign ministers' talks, Churchill commented disgustedly,

"They've bitched things up." That, at any rate, was what he said privately at Chartwell. He adopted a very different tone in a telegram to Eisenhower on July 17.

Churchill informed the president that he had made a great deal of progress physically, that he could now walk about, and that the doctors thought he might be well enough to appear in public in September. "Meanwhile, I am still conducting business." Indeed he was. Though the Washington communiqué had made no mention of leaders' talks to follow, Churchill craftily wrote as if Eisenhower had already accepted that the top men would come in later: "Please consider at your leisure whether it might not be better for the Four-Power Meeting to begin, as Salisbury urged, with a preliminary survey by the Heads of Governments of all our troubles in an informal spirit. I am sure that gives a much better chance than if we only come in after a vast new network of detail has been erected."

While Churchill waited for Eisenhower's reply, he confronted the problem of how to present the Washington agreement to the House of Commons. Obviously, there would be much disappointment with the mission's failure. Churchill was inclined to truthfully inform the House that Salisbury had agreed to foreign ministers' talks only after it became clear that the US and France would not accept the British proposal. The Foreign Office objected to this approach on the grounds that it would give offense to Washington and Paris. So, when Butler dined at Chartwell on the evening of July 19 to go over his speech, the wording he and the prime minister agreed on made a tortuous case for the supposed continuity between Churchill's May 11 address and the Washington communiqué.

Butler's speech on July 21 was a fiasco. The opposition bayed that in spite of anything the chancellor said, foreign ministers' talks were far from what Churchill had proposed. Attlee spoke of the great hopes that the May 11 speech had aroused and of the way in which the Washington talks had dashed those hopes by reverting to an approach that had failed in the past. The Labour member Ian Mikardo accused Butler and Salisbury of having sold out Churchill on behalf of a broader Conservative conspiracy against the prime minister. According to Mikardo, as soon as Churchill fell ill "the lesser men all round him got to work to tear down what he had been trying to build up. It was the mediocrities of the Conservative Party, the same people who had kept the Prime Minister out of office for all those years before the war, who seized the opportunity presented by his illness to betray those hopes."

In his absence, Salisbury came in for some particularly harsh personal abuse. Labour members gleefully mocked him as "limp" and untrustworthy, and claimed he had gone to Washington not to serve Churchill's Soviet initiative but rather to sink it. Part of this, naturally, was an effort by Labour, which had recently been weakened by party infighting of its own, to wreak similar havoc among the Tories by dividing Churchill from his party. But Labour also had reason to be suspicious of Salisbury. They knew of his opposition to the whole idea of top-level talks as articulated by Churchill. They had no inkling, however, of his and Butler's compact with Churchill or of both men's motives for faithfully doing the prime minister's bidding while they waited for Eden to regain his health. Contrary to Labour claims, Salisbury and Butler had made no effort to take advantage of the prime minister's illness. Ironically, it was Churchill who had used Eden's malady to neutralize the Lord President and the chancellor.

At Chartwell, Churchill was impatient to hear how Butler's address had been received. Unable to reach either Colville or Soames, he finally managed to talk to Buchan-Hepburn on the phone. Hardly had he received a preliminary account of the debate, in which both sides had vied to associate themselves with his summit proposal, when he heard back from Eisenhower, who fully and finally rejected it. Eisenhower pretended to have believed that the four-power meeting agreed to in Washington was consonant with what Churchill had wanted. He assured Churchill that it was all he would have consented to in Bermuda in any event. "I like to meet on a very informal basis with those whom I can trust as friends. That is why I was so glad at the prospect of a Bermuda meeting.

But it is a very different matter to meet informally with those who may use a meeting only to embarrass and entrap."

Salisbury came to Chartwell on July 23 to report on the Washington talks. He painted a vivid and disturbing picture. Since the 1952 presidential election campaign, the assumption in London had been that Dulles must be behind Eisenhower's more extreme foreign policy positions. Salisbury sought to correct that view. Based on what he had seen of the president in Washington, he now believed him to be the source of some of the administration's most dangerous ideas and attitudes. Eisenhower struck Salisbury as more "violently Russophobe" even than Dulles. The president spoke of harassing the Soviets "by every possible means" and of applying a policy of pin-pricks designed to bring down a Communist regime that Washington (though not London) suspected was already on the verge of collapse.

Salisbury and Churchill concurred that this ill-conceived strategy might have the quite unintended effect of setting off a third world war. But they differed sharply on how to follow up on Salisbury's sense that no argument of Churchill's would be capable of persuading Eisenhower to accompany him to the summit. Eisenhower opposed a leaders' meeting on constitutional grounds—the head of state must not argue. Salisbury believed that the danger from Washington made it more important than ever for the British to allow nothing to separate them from their American allies. As far as Salisbury was concerned, Churchill must altogether abandon the idea of seeing Malenkov on his own lest he put at risk Britain's ability to exert a beneficent restraining influence on the US. Churchill vehemently disagreed. For Churchill, the American danger made it more important than ever to get to the Soviet leader himself—before the world situation had a chance to ignite.

On the day of Salisbury's visit, reports in the French and American press quoted French Foreign Minister Bidault to the effect that though Churchill remained intact intellectually he suffered from complete paralysis, and that Eden was therefore urgently flying back from the US. It was true that Eden was due in London presently, but the reasons for his decision to put in an appearance before resuming his recuperation, on a Greek island cruise, were not as reported. In the six weeks since his third operation, Eden had had messages from well-wishers assuring him that Churchill meant to keep his place until October, that Salisbury was working to protect his interests, that Butler did not intend to move against him, and that all in all, his position was secure. Some supporters, however, were of the opinion that Eden would do well to show himself in London before Parliament rose for the summer recess or in early August at the latest. In line with this advice, Eden, who remained far from well, planned to stage a "triumphal" homecoming on July 26.

The stories that Churchill was unable to move without assistance so annoyed the prime minister that he spontaneously climbed up onto a chair and, as he stood erect with nothing to hold on to, challengingly asked his doctor what he thought of that. Reports that Eden was returning to take over from him prompted Churchill to try to tamp down the worst rumors about his condition. His wife had insisted that the staff at Chartwell must have some respite, so Churchill was set to transfer his base of operations to Chequers on the afternoon of Friday, July 24. The move provided an opportunity to be photographed in the doorway, as he prepared to leave for the three-hour drive from Kent to Buckinghamshire. The carefully staged image of the prime minister in a summer suit and bow tie, holding his hat in his right hand and cigar in his left, was his answer to suggestions that he was on his last legs and desperately in need of being replaced. The picture, which was the first the public had seen of Churchill since the stroke, ran in all the Saturday papers, on the eve of Eden's arrival. An accompanying statement from Number Ten announced that Churchill had "benefited greatly" from a month's rest and that at Chequers he expected to resume more of his normal work.

During this time, the Foreign Office was inundated with frantic telegrams from Eden. He expected to have radio and television coverage at the airport. He wanted photographers present when his plane touched down. These rumblings from America caused concern in both the Eden and Churchill camps that even now Eden might be expecting too much too soon. According to this view, for Eden's own good, he had to be made to understand that in spite of anything he had heard or been led to believe, Churchill was far from ready to lay down his burdens.

Evelyn Shuckburgh wrote to warn Eden that Churchill had begun to speak of plans that went well beyond October. Colville suggested that it would be best if Eden discreetly refrained from mentioning the succession when he and Churchill first met again. Norman Brook sent word that Churchill believed he had been doing Eden a favor by protecting his place and that he would not like to be asked when he could be expected to die.

Churchill was not the only one who was irritated by the timing and circumstances of Eden's return. Eden had signaled the Foreign Office that he intended to speak out about both the Washington talks and the May 11 speech when he faced the press at the airport on Sunday. At a moment when, as a consequence of the foreign affairs debate in the House of Commons, both topics had provoked a public uproar, Salisbury was not eager for Eden to weigh in and perhaps complicate matters. With a week to go before Salisbury had a chance to defend himself in the House of Lords, he urged Eden to keep his opinions to himself for the moment.

Salisbury was waiting at Heathrow Airport when the Edens flew in. As cameras clicked on all sides, Eden's wasted appearance mocked Salisbury's hopes that he would be well enough to replace Churchill in October. A thick coating of Newport tan could not conceal that Eden was a good twenty pounds below his normal weight. Churchill had done more than merely upstage Eden. In the duel of the newspaper photographs, the prime minister, who looked the healthier by far, was the clear winner.

When Eden came to Chequers with his wife for lunch the following day, the "two invalids" (as Colville dubbed them) eyed each other anxiously. Eden was predisposed to discover that Churchill could not last long in office, and Churchill in turn was eager for any sign that Eden was not ready to replace him. Eden's first sight of his host, sporting an enormous Stetson hat and making his way across the lawn unaided, was not encouraging. Throughout the meal, Churchill, who still spoke thickly at times and whose mouth remained slightly crooked, taunted Eden. "What!" Churchill exclaimed when Eden refused a nip. "Can't you drink? I can." At another point, Churchill assured his niece sympathetically, "Anthony must get his strength back."

Churchill had vowed beforehand that should Eden bring up the succession, he would say only that the more he was hustled the longer he would be. Churchill need not have worried—yet. On this first visit, Eden spoke largely of foreign affairs, of Beria's downfall and related subjects. But Eden and his wife, along with Lord and Lady Salisbury and others, were also invited to spend the following long weekend. In spite of further warnings from Butler, Buchan-Hepburn, and Bracken, Eden let it be known to his supporters that he planned to use the occasion to beard Churchill about the hand-over of power.

On the night of Tuesday, July 28, Salisbury and Shuckburgh discussed the coming collision as they worked late at Salisbury's London home. For days, Salisbury had been preparing the speech he planned to give in the House of Lords on Wednesday. This was the final run-through. Salisbury was known for his quality of reserve, his distaste for all emotional display. He could not, however, fully mask his agitation over the savage personal attack Labour had mounted against him. Unhappy about finding himself at the center of public attention, pessimistic about the chances of Eden's recovering in time, and convinced that his own recent exertions might have been for naught, Salisbury was in no mood to pretend yet again that Eden was up to confronting Churchill.

"I have been through all this a hundred times," Salisbury told Shuckburgh. "The fact is that the PM is much tougher than Anthony. He very soon brings Anthony to the point beyond which he knows he will not go and then he has won the day." It was a measure of Salisbury's exasperation that when Shuckburgh said he feared Eden would never get the premiership after all, Salisbury retorted that perhaps Eden's real role in life was to be a great foreign secretary—this after Salisbury had been scrambling for a month to finally install him at Number Ten.

Salisbury's performance the next day was pitch-perfect. Of the attack on him in the Commons he fairly dripped with scorn. "I noted with some interest that I was described as a kind of Jekyll and Hyde, a kind of schizophrenic, at the same time feeble, weak, limp—'limp' I think was the word—and deep, dark and dangerous. Personally I do not want to go into that. I am quite content to leave my character in the hands of your Lordships." Salisbury did however propose to answer head-on Labour charges that he had sold out Churchill. Mindful of the difficulties the Foreign Office had caused Butler when he was prohibited from attributing the failure of the British initiative to France and the US, Salisbury had previously secured Cabinet permission to place the blame squarely where it belonged.

In the course of his self-defense, Salisbury noted accurately that he had made the case in Washington for top-level talks with the Soviets. But, he went on, the French and the Americans had stood in the way. In short, the initiative had stalled not because of anything he had done but because of the obstructionist tactics of Britain's allies. After the speech, Salisbury was lauded in various quarters for the quintessentially Cecilian forthrightness of his remarks. Such praise had an edge of unintended irony given the circumstances that had led him to agree to travel to Washington in the first place.

As the fateful house party at Chequers drew near, Churchill had a conversation with the American ambassador, Winthrop Aldrich, that seemed to provide the opening he had been hoping for. Over lunch on Friday, July 31, Aldrich appeared to confirm Eisenhower's suggestion that in view of Churchill's health Eisenhower was prepared to come to him in London. Suddenly, Churchill saw a way around Eisenhower's unwillingness to risk attending a parley at the summit. While the president was in London, Churchill would seek his authorization to see Malenkov on his own.

By the time the Edens returned to Chequers on Saturday, their host was racing to orchestrate Eisenhower's invitation. The Salisburys were not due until teatime on Sunday. Eden, oblivious to what the prime minister was about to attempt, had decided to postpone his climactic face-off with Churchill until the reinforcements arrived. The atmosphere at Chequers was fraught. At Sunday lunch, Churchill and Eden clashed over the wisdom of meeting the Soviets. Clementine Churchill sided with Eden, though obviously for personal rather than policy reasons. She did not want Winston to risk a trip that his doctors feared could cost him his life. Churchill and Eden also differed in their views of Eisenhower, of whom Churchill declared, "He is a nice man, but a fool."

That evening, Churchill slipped off to the Royal Lodge at Windsor, a half-hour's drive, for his first audience with the Queen since his stroke. In their host's absence, Eden had a confidential talk with Salisbury, who had finally arrived at Chequers. Previously, Salisbury had doubted that anything would come of Eden's promised confrontation with Churchill. Now, he discovered that Eden no longer planned to broach the subject of the handover. He wanted Salisbury to do it for him. Salisbury agreed but suggested that they had better sleep on it.

Presently, the question of whether or not to press Eden's claim vanished from consideration, when an ebullient Churchill returned from Windsor. He had managed to secure the Queen's permission to ask Eisenhower for the last week in September or the first week in October. The Queen promised to return from Balmoral to welcome Eisenhower at Buckingham Palace during his stay. When Salisbury learned of Churchill's scheme to seek the president's blessing to meet Malenkov he was horrified. Bilateral talks with the Soviets were precisely what Salisbury had warned Churchill against on his return from Washington. To add insult to injury, Churchill had enlisted the Queen's support without so much as a word to Salisbury or other colleagues beforehand. Salisbury objected immediately, but it was too late. Churchill presented the matter of the state visit as a fait accompli. An invitation flew off to Washington the next day and Churchill gaily bid farewell to his houseguests without their having had a chance to quiz him about his retirement plans.

All of Churchill's fighting instincts were in play, and when Violet Bonham Carter visited two days later, Clementine Churchill, who arranged to see her first, implored her not to encourage him to carry on. Clementine explained that though her husband's mind was clear, he tired easily, and she was certain he ought to give up power in the autumn rather than wait and peter out. When she saw him, Violet realized that in spite of anything she or anyone else might recommend he did not intend to go gently into that good night. After the visit, she wrote in her diary of the man she had known and adored since she was nineteen and he thirty-two, and whom she had once hoped to marry: "I drove back feeling an unutterable sense of tragedy—at watching this last—great—ultimately losing fight against mortality. The light still burning—flashing—in its battered framework—the indomitable desire to live & act still militant & intact. Mind & will at bay with matter. At best a delaying action—but every instinct armed to fight it out until the end."

Churchill was not altogether unrealistic about whether his body would permit him to continue in office. He told Violet Bonham Carter, as he had previously informed the Queen, that before he took a final decision he must see if he could clear two hurdles. As far as he was concerned, he had to be able both to deliver the leader's speech at the Conservative Party conference and not only to face, but also to dominate the House of Commons. He really did not know if he could manage to bring off either. When Eden left England on August 8 to resume his own recuperation in peace and tranquility, Churchill plunged headlong into a series of what he viewed as preliminary tests. On the day of Eden's departure, Churchill presided over his first official meeting since he faced the Cabinet on the morning after the Gasperi dinner. At Chequers, he conferred for an hour with Salisbury, Butler, and William Strang of the Foreign Office.

Churchill brought the meeting off to perfection, and was in a "sparkling mood" at the small luncheon that followed, where they were joined by some others. In the course of the meal, the conversation turned to Churchill's prose style, and Butler asked whether Macaulay had been one of his literary influences. Churchill said he had, as well as Gibbon—whereupon he launched into a recitation from memory of the opening lines of Chapter Fourteen of *The Decline and Fall of the Roman Empire*, about the eighteen years of discord and confusion that afflicted the empire after the abdication of Diocletian: "The balance of power established by Diocletian subsisted no longer than while it was sustained by the firm and dexterous hand of the founder. It required such a fortunate mixture of different tempers and abilities as could scarcely be found, or even expected, a second time."

As the moment approached when Churchill would be expected to hand over, this was his puckish way of reminding his two chief co-conspirators that it might not be in the country's best interests to lose him. No one savored the performance more than Salisbury. Nonetheless, Salisbury persisted in the belief that he had never known a more dangerous situation. He insisted afterward to Macmillan (who had not attended the luncheon) that while Churchill was "more charming and delightful than ever," he simply had to resign as soon as Eden was ready.

Two days later, Churchill's sky darkened. Eisenhower would not be coming to London after all. He said he was afraid that Aldrich had misinterpreted a little wishful thinking on his part, growing out of disappointment at not being able to see Churchill in Bermuda; or perhaps Churchill had misunderstood Aldrich. As it was, "a number of inescapable commitments during the foreseeable future" made it impossible for Eisenhower to leave the country. Clementine Churchill did not believe a word of it. She was sure that in fact Eisenhower did not wish to see her husband and that he had been delighted when Bermuda had had to be canceled. She guessed that at the moment, Eisenhower was merely waiting to see which way the cat jumped and whether the prime minister would retire. Churchill briefly considered proposing that he come to Washington in late September. But he soon abandoned the idea, at least until he had convinced himself that he had the stamina to press on.

Against his doctor's advice, Churchill presided over his first Cabinet, in London on August 18. Though he had been uneasy about it, the session seemed to go well, and he dined with friends late that night. But when he arrived at Chartwell the next day, he looked alarmingly "dazed and gray." His family was dismayed by what the meeting had cost him, and he wondered aloud about his ability to tackle the party conference when a single Cabinet had been capable of depleting him so severely. The incident set off a new wave of uncertainty. In the days that followed, Churchill spoke by turns of handing over in October and of postponing his departure until May 1954 when the Queen returned from her round-the-world Commonwealth tour. He alternated between believing that he was too weak to fight on, and telling himself that it was his duty to keep his Soviet initiative alive when the Americans were so opposed to it and when Eden, if prime minister, was certain to side with Eisenhower. Churchill's family had difficulty believing he would yet be able to recover sufficiently to return to public life. He accused them of selfishness in urging him to resign when he retained the capacity to help the country.

Still, he did not wish to proceed recklessly. He did not want to attempt either the party conference or Parliament if he believed he was not up to the task. Certainly, Brain, the neurologist, was pessimistic. A great many things could go disastrously wrong at Margate, where the party conference was to be held. The stroke had diminished Churchill's control over his emotions, so he might embarrass himself on the platform. It had robbed him of strength, so he might grow exhausted and have to make an ungraceful exit. It had affected his memory, so he might forget what he had come to say. Brain also doubted that Churchill would be in any condition to return to the House of Commons. Churchill decided to wait and see how he felt in September.

When Queen Elizabeth invited Churchill to accompany her to the races at Doncaster on September 12 and to spend several days afterward with her and Prince Philip at Balmoral, he accepted despite his wife's vehement objections. Clementine Churchill worried that Doncaster, which would require him to stand and walk in front of affectionate but curious crowds, would be too much for him. Nor was she pleased to see him subject himself to the strain of an overnight trip in the royal train to Scotland. As the outing approached, Clementine's anxieties mounted. During this time, she herself suffered a bad fall and cracked her ribs. On September 2, she was in London to see her doctor, when she telephoned Winston at Chequers to urge him again to send his regrets to the Queen. Winston dug in and the couple quarreled bitterly. Before the evening ended, however, he called her back full of apologies and tender words.

The morning after, Clementine composed a note in the same loving spirit. She warned him not to risk a setback before Margate and Parliament. He, however, would not be deterred. In the end, Clementine joined him at the races—which by chance took place on their forty-fifth wedding anniversary—and at Balmoral. He did exceedingly well at Doncaster; and in Scotland, rather than permit himself to be treated as an invalid, he insisted on trudging at the Queen's side through the heather for some three-quarters of a mile. He seemed a good deal more confident when he returned from Scotland, yet there could be no denying that, much as Clemmie had feared, the expedition had overtaxed him.

On September 17, Churchill went off for two weeks to La Capponcina, during which time he would take a decision about the party conference. From the moment he arrived it was evident that his energy and stamina were, as he told his daughter Mary, who accompanied him, on an ebb tide. He had hoped to paint but found that he was unable to marshal the strength even to go outdoors and sit at his easel. He was often depressed, and though he persisted in working on a speech for Margate he questioned how long he could stand on his "pins" to deliver it. "I still ponder on the future," he wrote to Clemmie, "and don't want to decide unless I am convinced." Finally, he did manage a bit of painting. Working on a picture was a great relief, he reported, "and a little perch for a tired bird." Churchill's agonizing indecision persisted for the better part of his stay, but by the time he flew home on September 30 he had made up his mind to try both hurdles. He promised himself that if his body failed him during either ordeal, he would step aside.

Margate loomed in ten days. First, however, he had to handle Eden, who flew in from Athens shortly after the prime minister returned from the South of France. During his sojourn in Greece, Eden had had ample warning, from Salisbury and Macmillan among others, of what faced him. Nevertheless, at the airport he pointedly avoided telling reporters that he was headed back to the Foreign Office. As if to leave open the possibility that the prize might yet be his, Eden would say only that he was "fit and ready to work." A session with Churchill the morning after his return swiftly shattered any lingering illusions he might have harbored. When Churchill and Eden talked on October 1, the prime minister left no doubt that he wanted to stay put. He aimed to try himself out and he promised Eden that if he found he could not do his duty he would resign.

Churchill asked Eden to return to his old post as foreign secretary. Eden worried that if he declined, it would look as if he were not fit and that that might be a cue for Butler or one of the other possibles to swoop in and take his place at the head of the list. Still, the meeting concluded without his having answered Churchill one way or the other.

That evening, Eden, Salisbury, and Butler were the prime minister's dinner guests. In front of the two men who had done the most to keep the premiership open for him, Eden caved in to Churchill yet again. He announced that he would do as Churchill requested and resume his duties as foreign secretary. In short, whatever Salisbury and Butler may have thought they were accomplishing at the time, the sole outcome of the medical cover-up had been to allow the old man to cling to office long enough to engineer another comeback. In the guise of catering to their interests, Churchill had made fools of them both.

Tonight, he rubbed their noses in their deeds when he announced that now that he was back he meant to resume his pursuit of a meeting with Malenkov—with or without Eisenhower. Tempers flared at the disclosure that Churchill planned to publicly revive his May 11 proposal when he addressed the Conservative Party conference. Eden made it clear that he did not approve of the May 11 speech. When Churchill huffed that it had been very popular, Eden shot back that he did not dispute that. Popularity was not the point; he opposed Churchill's policy regardless of public opinion.

When Eden and Salisbury joined forces to question the usefulness of top-level talks, Churchill jested darkly, "Now I begin to understand what Chamberlain felt like." The joke outraged Salisbury, who flung back that this was just too much. At one point, Churchill and Eden debated whether the leaders' or the foreign secretaries' meetings had been the more important during the war. Churchill insisted that the Stalin-Churchill-Roosevelt meetings had made the foreign secretaries' work possible. Eden protested that it was his own discussions with American Secretary of State Cordell Hull in Moscow in 1943 that had been decisive. Nothing was resolved and the evening ended with Churchill and his guests at daggers drawn.

Eden returned to work at the Foreign Office on October 5 determined to find some way to stop Churchill from going to the summit. His instinct, as so often in their dealings, was to look for someone else to do the dirty work. Eden initially saw Eisenhower as the man for the job, and he considered pressing for an allies' meeting where Eisenhower could be counted on to torpedo top-level talks. But Eden soon grew worried that Churchill would use the announcement of a meeting with the president to build even more momentum and public support for a parley at the summit. An encounter with Dulles, which would draw less attention, might be another matter. Dulles had already suggested that he and Eden sit down to talk as soon as Eden completed his convalescence.

Churchill vetoed Eden's suggestion that Dulles be invited to

London. Suddenly, the prime minister was touting a different idea. Why should the two top men and their foreign secretaries not meet at the same time? Three days before Margate, Churchill sent off a message to Eisenhower proposing that they meet the following week in the Azores. While the leaders conferred, Dulles and Eden, who would accompany them on the trip, could meet as well. Having renewed his May 11 proposal at the party conference, Churchill would use the opportunity to ask Eisenhower either to join him at the summit or to authorize him to see Malenkov on his own. Either way, when Churchill returned to London he would be able to make his first appearance in the House of Commons since the stroke with his return ticket to the summit finally in hand.

Mindful that Eisenhower had declined his previous invitation, Churchill meant to force the issue this time. His October 7 telegram was less a request to meet than an offer to the president to choose the venue. Churchill reminded Eisenhower that he had not troubled him with telegrams since August. He said he would be agreeable if Eisenhower insisted on including the French, though Churchill for his part would prefer to do without them. The meeting, however, had to take place the following week, before the opening of Parliament. Churchill suggested that in view of his and Eden's health it would be easier to go to the Azores than to Washington. But, he added, if Eisenhower found that impossible he and Eden would be at the British embassy, Washington, during October 15–18. Churchill did not say so but the implication was clear. As a matter of courtesy, the president would have to see them then.

While Churchill waited to hear back from Eisenhower, he plunged into preparations for the party conference. Two days before he confronted the first of his self-imposed hurdles, he rehearsed his big speech exactly as he planned to deliver it. On the day of the run-through, he dined at noon: a dozen oysters, a bit of steak, and half a glass of champagne. An hour later, he swallowed the pill that Moran had specially concocted to clear his head and provide a blast of energy before he spoke. An hour after that, he stood on his "pins" and sped through the speech in a little over thirty-six minutes. Clementine Churchill sat behind him, precisely as she would at Margate. The practice session (not to mention the amphetamines) left Churchill feeling elated and confident. He planned to travel to Margate on Friday afternoon so he could spend the night there and be well rested.

On Friday morning, however, all hell broke loose. For weeks, Eisenhower had indeed been waiting to see if Churchill would retire when Eden came back. Now that it was apparent that Churchill meant to go on, the president's patience seemed depleted. Eisenhower icily refused to meet in the Azores. Nor would he agree to Washington, as he said he would be out of town during the dates mentioned. If Churchill and Eden insisted on coming anyway, the best that could be arranged was to have Dulles talk to them. If they could stay through October 20 (which Churchill had stressed they could not), Eisenhower would participate on that day, but not earlier. Though it was evident that Churchill's principal purpose was to confer king-to-king, Eisenhower suggested that it might be best to send Eden alone.

Eisenhower's answer infuriated Churchill. He decided to put Eisenhower on the spot by suggesting that he and Eden might come a week later, though that would require Churchill to miss the first week of the new parliamentary session. When Eden's Foreign Office colleagues urged him to find a way to stop Churchill from provoking Eisenhower, Eden replied gloomily, "I haven't got a log heavy enough to hold this elephant." Twice on Friday afternoon, Eden appeared at Number Ten with drafts of messages which reverted to Eden's original plan to invite Dulles to London. Twice, Churchill refused. Finally, the prime minister left for Margate, but hardly had he arrived at his hotel when he was on the phone to browbeat Eden into agreement that they try again to force Eisenhower to come to the Azores. Eden buckled and a new message to the president was drafted. This occasioned acute anxiety at the Foreign Office about what Churchill's next step might be should the president refuse again. Nevertheless, by 7 p.m. on Friday the telegram was ready to go and was sent off to be enciphered.

At the last minute, a remark of Colville's caused Churchill to reconsider. Colville asked, "What subjects are you going to discuss when you get there?" It occurred to Churchill that at this point anything he might say to Eisenhower face to face would necessarily elicit a negative response. Within a matter of hours, Churchill would have a platform where he could begin to apply intense public pressure on the president to be more agreeable. He therefore canceled the first telegram in favor of a softer, more conciliatory one that offered no hint of the fireworks to come. Churchill wrote that in view of Eisenhower's schedule it did not look as if the meeting he had hoped for could be arranged. "I am very sorry, as there are so many things I would like to talk over with you quietly and at leisure. I earnestly hope a chance may come in the not too distant future." In the meantime, he told Eisenhower, Eden would cable separately to ask Dulles to London.

Some four thousand Conservatives gave a standing ovation and sang "For He's a Jolly Good Fellow" when Churchill appeared on the flower-edged stage at the seaside Winter Gardens the next day. Despite the adoring reception, Churchill knew how much depended on the performance he put up. There had been rumors that he had suffered a stroke and rumors that he had been rendered unfit to lead. Aneurin Bevan had loudly demanded that Churchill "clear out" if he was no longer capable. *The Times* had called on the prime minister to demonstrate that he was physically equal to the tasks ahead.

Would Churchill's many well-wishers at the conference think that he seemed different? Would they notice the faint crookedness in his mouth, the residue of stiffness in his arm? Having shown him their love, would they soon decide to show him the door? When Churchill put on his glasses and began to read from notes, none of the ministers and party officials seated behind him knew, in the words of one, "whether he would pull through or not."

The speech was an ordeal, but the fact that the whole world was watching made it an opportunity as well. Hours after Eisenhower had cold-shouldered him, Churchill renewed not only his May 11 proposal, but also his bid for the diplomatic leadership of the West.

He announced that the reason there still had not been a parley at the summit was that Britain's "trusted allies" had yet to agree to the idea. Churchill put those allies on notice that he meant to keep pounding until he prevailed. That meant retirement was out of the question—at least for now.

"If I stay on for the time being, bearing the burdens at my age," Churchill continued, "it is not because of love for power or office. I have had an ample feast of both. If I stay, it is because I have the feeling that I may, through things that happened, have an influence on what I care about above all else, the building of a sure and lasting peace."

When Churchill concluded, the hall exploded in applause. All he had been required to do was show that he was in fighting trim. No one had expected him to come out swinging. *The Times* later called his performance a "triumphant return to public life" and the consensus in the party was that Winston was back—again. After the young MP Edward Heath moved a vote of thanks to "the great Prime Minister of peace," a visibly weary Churchill began to leave the platform on his way to an adjoining room, where an overflow crowd had listened to his speech on loudspeakers. He was scheduled to say a few words to them, but suddenly he realized that any further exertion was beyond his powers.

Churchill reappeared at the microphone to say, "I hope you will excuse me. I have done quite a lot of work today. My trusted colleagues Mr. Eden and Lord Salisbury will take my place."

Yet again, Churchill had upped the ante in his long-running struggle with Eisenhower by appealing directly to the people—not just Britons, but people everywhere—to support his initiative. Eisenhower reacted by issuing strict marching orders to Dulles on the eve of his departure for London. He assigned Dulles to put an end to Churchill's summit appeals by making it clear once and for all that the president would never agree. Dulles was to leave no doubt in Churchill's mind that coming in at the very last minute to finalize a deal that the foreign ministers had already negotiated was as far as Eisenhower would ever go.

Churchill's remarks at the party conference had angered and irritated Eisenhower, but they had also made him exceedingly nervous about what the prime minister might have accomplished by taking his case directly to the people. It was one thing to privately resist Churchill's arts and quite another to refuse him with the whole world looking on. Churchill had highlighted the strategic breach between London and Washington, and Eisenhower worried that people would look badly on US opposition to the prime minister's proposal. So, even as he directed Dulles to decline any suggestion of leaders' talks with the Soviets, he acknowledged that the secretary of state "would be walking a tightrope in maintaining this position without creating the impression that the US was blocking a useful step."

Dulles, whose breath stank and whose left eye twitched incessantly and disconcertingly, arrived in London on October 15. That night, he dined at Number Ten, where Churchill's other guests were Salisbury and Eden. When the prime minister spoke of his summit proposal, Dulles moved at once to try to turn Churchill from his purpose. Churchill responded to Dulles's invocation of the constitutional argument by politely conceding that the president's position differed from his own. Then he added that, as Eisenhower felt he could not sit down with Malenkov, perhaps he, Churchill, ought to go alone. In a similarly courteous tone, Dulles observed that obviously this was a decision Churchill was wholly free to make. But, Dulles added, he was concerned that a solo mission would create the impression that Britain had assumed the role of middleman between America and the Soviet Union. He suggested that such a perception would "seriously prejudice" America's desire to work in close partnership with Britain in areas of mutual concern.

Suddenly, the air of cordiality that had prevailed thus far in the encounter dissolved. Churchill retorted angrily that he thought the Americans could trust him not to be entrapped at Moscow. He recalled that when he met Eisenhower in New York prior to the inauguration, the president-elect had said he was free to meet the Soviets alone. Dulles in turn assured Churchill that he was certain this remained Eisenhower's view. Of course, Washington would make no effort to interfere in any decision Churchill might make. Still Dulles felt obligated to point out that as Churchill could hardly go as Washington's representative, US public opinion would inevitably cast him in the role of middleman and that that could have an undesirable effect on relations elsewhere. Dulles had just put on the table the very possibility that Salisbury, arguing from a dramatically different perspective, had warned of when he returned from Washington—that Churchill's summit strategy risked the Anglo-American alliance. When in the course of the dinner Dulles interpreted Salisbury's silence to mean that he had the Lord President's tacit support, he was right, but only up to a point. Salisbury and Dulles differed about much else, but for the moment their shared opposition to bilateral talks made them strange political bedfellows.

Eden too had remained significantly silent throughout Churchill's acrimonious exchange with the American visitor. Dulles had gone into the dinner with a comforting sense that as far as the summit debate was concerned he and Eden were on the same side. During the drive from the airport, Eden had conveyed to Dulles that though he would of course "loyally" support his chief, he personally doubted the wisdom of top-level talks. When in the days that followed, Eden, Dulles, and Bidault sat down together, the foreign ministers, clearly operating on a very different line from Churchill, agreed to renew the Western allies' stalled offer to the Soviet Union of a second-level, four-power conference on Germany and Austria. A communiqué to announce the invitation was drafted and on October 18 Churchill hosted a farewell luncheon, where it quickly became apparent that American efforts to silence him had failed. When he reviewed the draft communiqué, Churchill immediately complained of the wording. As had been the case before, he knew he could hardly block a foreign ministers' meeting, but he did not wish the communiqué to undermine his goal of a parley at the summit. He objected strongly to the statement that the three governments agreed that a foreign ministers' meeting would be "the most practical step toward a reduction of international tension and a solution of major European problems," and urged that the communiqué say instead that as far as their governments were concerned, a foreign ministers' meeting "might be an invaluable step."

Things heated up from there. Churchill went on to make his usual derogatory remarks about the usefulness of foreign ministers' talks. To the fascination of Churchill's guests, his comments provoked an outburst of annoyance from Eden. On October 1, the two men had sparred privately and indeed rather bitterly on the subject. The difference on the present occasion was that they were openly disagreeing in front of the Americans and the French. Salisbury dove in to break up the fight, but by that time, the damage to the premiership had been done. Dulles now had no doubt that the prime minister was in this alone. On the matter of a summit meeting, Churchill perceptibly did not have the backing of his own key people.

Something else happened at the luncheon that raised eyebrows on the American side. After the altercation with Eden, Churchill seemed suddenly in much less good form than at the start of the meal. Aldrich, who was among the guests, gauged that Churchill was finding it hard to concentrate. At times, the prime minister's comments sounded suspiciously like a set speech and did not really fit into the discussion. Apart from various individuals' particular reasons for opposing a parley at the summit, episodes like these raised important questions about what could happen if Churchill succeeded in his campaign to face the Soviets across a conference table.

There was no doubt that he retained the ability to soar intellectually and strategically. But, seemingly more and more, there were also moments when he crashed, when he was simply too groggy to operate effectively. These days, one never knew which Churchill one was going to encounter at any given moment. Even if one agreed with him on the desirability of leaders' talks, the question unavoidably presented itself of whether it would be responsible to allow the old man to attend.

For the moment, it seemed as if foreign ministers' talks was the route the Western allies had chosen to go. In the days that followed, however, Churchill made it clear to the world that he had other ideas. It was as if he had not heard a word that Dulles had said in

London, as if he had not been firmly warned of the consequences should he persist in badgering Eisenhower. On October 20 during Prime Minister's Questions in the House of Commons, Churchill stressed that in spite of the foreign ministers' invitation, leaders' talks remained his goal; and again on October 27 he suggested in the Commons that he longed to have a private talk with Eisenhower, presumably to discuss the next step.

During this time, the thirty-part newspaper serialization of *Triumph and Tragedy*, the concluding volume of Churchill's war memoirs, which had begun to run on October 23, ratcheted up the public pressure on Eisenhower by richly documenting Churchill's claims that he had seen the Soviet danger in 1945 but that the Americans had failed to listen. The passages about Eisenhower had been toned down somewhat in view of his current employment, but enough of the sting remained for the *New York Times* to have asked the president in advance of publication whether he wished to respond. Eisenhower had long and accurately maintained that he had not been ordered to take Berlin. But after Churchill published extracts of their correspondence, it could never be said that Eisenhower had not been warned of the political and psychological importance of beating the Russians to Berlin, or that he had not been cautioned that it would be a huge mistake to put his faith in Stalin. Once again, the British prime minister was openly battling Eisenhower on two fronts, the past and the present, for the verdict of history and for the diplomatic leadership of the West. This time, his documentation provided him with especially powerful ammunition. The fact that in October Churchill was awarded the Nobel Prize for Literature heightened the excitement surrounding his memoir.

The serialization was in progress when, on November 3 Churchill used the occasion of the culminating hurdle, his first full-dress parliamentary speech since the stroke, to again call for a parley at the summit. Churchill alluded movingly to his own advanced age when he emphasized that at this point, it probably was no longer possible to obtain a speedy settlement: "Time will undoubtedly be needed—more time than some of us here are likely to see." Churchill hoped at least to have the opportunity to begin the process. But before he could even try, he had to wrest some form of assent from the White House.

Two days after his address, when Moscow responded negatively to the foreign ministers' invitation, Churchill saw a chance to appeal to Eisenhower anew. This time his proposal took the form of a private communication, but he held open the possibility that he might be forced to speak out again if he did not get what he wanted. "The Soviet answer puts us back to where we were when Bermuda broke down through my misfortune," he wrote to the president on November 5. "We are confronted with a deadlock. So why not let us try Bermuda again?" Pointedly and not a little threateningly, Churchill warned of "serious criticism" from the public if a meeting could not be arranged.

The following day, Eisenhower capitulated. He unhappily agreed to face the prime minister during the first week of December, with the proviso that their talks must not be billed as the prelude to a further approach to the Soviets. As he had done six months before, Churchill offered hand-on-heart assurances that he would suggest no such thing.

Churchill had had no help from the Cabinet in this latest struggle with Eisenhower. Now that he had won, he was hardly in the mood to listen to ministers' advice about his travel plans. Churchill rejected the argument that in view of his health it would be safest to travel by sea. "I shall go to the airport," he told them. "I shall board my plane. I shall go to bed and take my pill. In the morning I shall wake up either in Bermuda or in Heaven—unless you gentlemen have some other destination in mind for me."

XVII

I HAVE A RIGHT TO BE HEARD!

Bermuda, December 1953

Asked who exactly was standing in his way right now, the seventy-nine-year-old Churchill shot back, "Ike."

At dinner with his team at the Mid-Ocean Club in Bermuda on the eve of Eisenhower's arrival, Churchill had been speaking of his great plans. The goal was to get to Malenkov. To accomplish that, he had to persuade Eisenhower either to accompany him to the summit or to authorize a solo mission. It was a tall order, to be sure. Still, against all odds Churchill had managed to get this far since his stroke, and for the most part he seemed radiantly (certain of his colleagues thought unrealistically) confident that face to face he could get what he needed from the president.

When Eisenhower flew in on December 4, 1953, it was no secret that he was a reluctant participant in the conference. This was to be his first encounter with Churchill since the May 11 speech and the publication of *Triumph and Tragedy*, in both of which the prime minister had posed a controversial and highly effective public challenge to the president's authority. Churchill's ability to keep returning from the dead (political and otherwise) had long been maddening to his antagonists, and so it was now to Eisenhower. Yet again in the Churchill story, the funeral had been premature, the coffin empty. Eisenhower was furious that Churchill had ignored Dulles's warnings and had persisted in pressuring him to meet. In the weeks since Bermuda had been agreed to, Eisenhower had sought repeatedly to douse press expectations of what might be accomplished there. When Eisenhower emphasized that the whole idea had been Churchill's, when he predicted that no real news would emerge from their discussions, and when he stressed that he had consented merely because he wanted to be agreeable, he made the conference sound like nothing more than an old man's folly.

Eisenhower looked anything but agreeable when he emerged from his aircraft with a grin that struck observers as forced. After he greeted Churchill, he strode off to inspect an honor guard of Welsh Fusiliers. Churchill was left "trotting behind, neglected and pathetic." The prime minister refused to be offended and finally "pirated" Eisenhower and arranged a tête-à-tête lunch in his suite.

At half past one, when he finally had Eisenhower to himself, Churchill promptly broached the possibility of a meeting with the Soviets. Since the Bermuda conference had been arranged, Moscow had reversed its earlier rejection of foreign ministers' talks. As a consequence, part of its purpose had become to draft the West's reply to the most recent Soviet note. When Churchill immediately spoke of keeping the door open to top-level discussions, Eisenhower was just as quick to cut him off. He reiterated what he had previously sent Dulles to London to convey: he would not participate in a leaders' meeting until Moscow had shown good faith at the foreign ministers' level. If Churchill thought he could continue to badger him in the press, Eisenhower had a surprise waiting.

When the American, British, and French delegations convened that evening, the president disclosed that he had received an invitation to address the United Nations. Though he emphasized that he had yet to accept, he reported that he had already drafted a speech on the peaceful use of atomic energy. His address would present the US as it really was—"struggling for peace, not showing belligerence and truculence, but rather our will for peace." When Churchill learned that Eisenhower was to deliver a major address on the same evening he left Bermuda, December 8, he sensed that the president had outflanked him. Obviously, Churchill would be unable to stop him from giving the speech, the latest blow in their contest for preeminence. Whatever the outcome in Bermuda, Eisenhower's UN appearance would overshadow the allies' meeting and dominate the news. Should Eisenhower persist in blocking a parley at the summit, the New York speech would armor him, at least for the moment, against further charges by Churchill.

Nevertheless, following some opening remarks by French prime minister Joseph Laniel, Churchill methodically laid out the strategy that underpinned his desire to meet Malenkov. He spoke of a "new look" in Soviet policy, the apparent change of heart that had occurred since Stalin's death. He proposed that the West adopt a double-barreled policy of military strength combined with negotiations, and he suggested that it might be possible to open up the closed Communist society by contacts, meetings, trade, and other forms of "infiltration" behind the iron curtain. Churchill argued that such infiltration was feared by everything that was bad in the Kremlin regime. Western leaders, by contrast, had nothing to fear from top-level talks.

Eisenhower sat across from Churchill at a large, round, cedarwood table that, due to a power failure, was illuminated by candles and hurricane lamps. He gave the prime minister all the time he needed to frame his argument. But when Churchill was finished, Eisenhower stunned the British delegation with a smackdown of a speech that witnesses characterized as "vulgar," "rude," and "very violent." In his diary afterward, Colville expressed doubt that such coarse language had ever before been heard at an international conference. Eisenhower compared Russia to a woman of the streets and he insisted that whether her dress was new or just the old one patched she was still the same whore underneath. He said it was America's intention to force her off her current beat into the back streets.

Eisenhower's rant provoked "pained looks all round." Eden had hoped that the president might be the man to stop Churchill. He shared Eisenhower's antipathy to leaders' talks, as well as his belief that Moscow's somewhat more reasonable attitude lately was attributable less to Stalin's absence than to constant pressure from the West. He also shared Eisenhower's ineffable frustration at Churchill's refusal to vanish into retirement and at the persistence, bordering on effrontery, with which the old man had forced this conference to take place. Still, the tenor of Eisenhower's remarks rattled him. In an effort to wind up the proceedings on a note of dignity and decorum, Eden leaped in to ask when the next meeting would be.

"I don't know," Eisenhower snapped, before he rose to leave. "Mine is with a whisky and soda."

The rancorous session set the tone for a conference that would be marred by frayed tempers on both the American and British sides. (The French amused themselves by leaking choice bits to the press.) The next day, when Churchill and Eisenhower squared off on atomic policy, it seemed to Churchill that all else shrank in significance alongside Eisenhower's declared determination to unleash atomic weapons on Chinese bases should the truce in Korea break down. China did not have the bomb but the Soviets, who did, were pledged to come to their Communist ally's defense, and Churchill saw Britain as a likely first target.

Churchill warned the president that the use of atomic weapons in Korea could trigger a third world war. Throbbing with emotion, he explained that the defenses in London were inadequate against such weapons and that he could not bear to think of "the destruction of all we hold dear, ourselves, our families and our treasures." He went on, "Even if some of us temporarily survive in some deep cellar under mounds of flaming and contaminated rubble there will be nothing to do but take a pill to end it all." Eisenhower's response was to scoff at the notion that the use of atomic weapons in Asia would provoke the Soviet Union to attack the West.

Churchill despairingly produced a copy of the Quebec Agreement of 1943, signed by himself and Roosevelt. That agreement had given Britain veto power over the use of the bomb. Had it still been in effect ten years later, Churchill could simply have nixed the whole idea of strategic bombing and Eisenhower would have had little choice but to back off. As it was, the Attlee government had abandoned the British veto, so all Churchill could do now was plead.

On this and related matters, he pleaded in vain. After his stroke, Churchill had pulled himself back from the brink of death. He had persisted in fighting and scheming when other men might have long given up. Yet, in the end, he had failed with Eisenhower in every significant respect. The president had agreed neither to meet Malenkov nor to give his blessing to Anglo-Soviet talks, and he had not backed down about using atomic weapons in Asia. The sole positive outcome of the conference was to begin preparations for another foreign ministers' meeting with the Soviets, this one to take place in Berlin in January.

Churchill bitterly attributed his failure in Bermuda to age and incapacity. He began to accept what his doctor had known for some time, that no matter how hard he pushed he would never again be the man he was before the stroke. Reflecting on his dashed hopes for the conference, Churchill tearfully told Moran, "I have been humiliated by my own decay."

In London, the excitement and optimism that had surrounded Churchill's departure for Bermuda had fizzled out by the time he came home empty-handed. To his family and close associates he spoke often of retirement. No one really expected him to lead the party into a new general election, which had to be held by the fall of 1956. (British parliaments are elected for five years, but the party in power can call a new election at any point.) It was tacitly understood that Churchill would not want to risk ending his career with another defeat. The hope among party leaders was that he would retire decently in advance of the next election in order to give Eden time to establish himself. Churchill went so far as to mention various dates when he thought he might like to go. Colville conveyed to Shuckburgh that the prime minister would be much more likely to relinquish power if Eden was "kind" to him and that every time

Eden acted too aggressively Churchill saw it as evidence that he was still not ready to be left in charge.

Eden for his part complained to colleagues that Churchill was "gaga" and that the situation must not be allowed to go on. There was no question that Churchill was suddenly much deafer, that he repeated himself incessantly, and that he conducted Cabinet business at a glacial pace. Slumped in his seat, with his shirtfront gathered in a large roll, he seemed robotic at times and appeared not to recognize people he ought to have known. In the House of Commons, he made a show of laughing off a Labour backbencher's questions about when he might be expected to stand down. Yet he could not disguise his hurt when *Punch* ran a controversial cartoon that cruelly suggested the effects of the stroke on his face. An accompanying text penned by the editor, Malcolm Muggeridge, accused Churchill of clinging to office with faltering capacities. In this increasingly ugly and harassing environment, Christopher Soames, privately acknowledging that his father-in-law had deteriorated significantly during the first few weeks of 1954, concluded that it should be Churchill's friends rather than his adversaries who convinced him to go.

There was little chance that Churchill would agree to do any such thing after a strange and disquieting message arrived from Washington on February 9. Suggesting that he and Churchill shared a responsibility to protect civilization against the incursions of "atheistic materialism," Eisenhower spoke of the need to seek renewed faith and strength from God and to sharpen their swords "for the struggle that cannot possibly be escaped." What did Eisenhower mean by that last phrase? Was he talking (as Churchill hoped he was) about a spiritual struggle with the Soviets—or did he mean inevitable war? Salisbury, when at length he saw what Eisenhower had written, was if possible even more alarmed.

Churchill had yet to make up his mind about quite how to reply when a newspaper report of a speech by the chairman of the Joint Congressional Committee on Atomic Energy made the eyes "start out" of his head. Addressing a group of businessmen in Chicago,

Representative W. Sterling Cole offered the first extensive public account of a US hydrogen bomb test eighteen months previously that had obliterated a Pacific island and torn a vast cavity in the ocean floor. The congressman noted that the Soviet Union was also engaged in H-bomb research and was perhaps a year behind the Americans. When Churchill contacted the White House on March 9, he had a simple, powerful new argument. He believed that the Cole revelations had shifted and narrowed the topic of the conversations he and Eisenhower needed to have with Malenkov. To Churchill's mind, the Soviet leadership must be made to understand that the hydrogen bomb, with its vast range of annihilation and even broader area of contamination, rendered the scattered population of the Soviet Union vulnerable as never before.

Churchill was waiting to hear back from Eisenhower when, on March 17, the Soviet chargé d'affaires in London transmitted a cryptic message to the effect that Malenkov "would be willing to consider meeting him." Churchill took this to mean that Malenkov was interested in talking to him alone. When he reported the development to Eden, the foreign secretary was most discouraging. The Berlin foreign ministers' conference (January 25–February 19) had produced no results, but before it ended Eden had succeeded in the delicate task of getting all parties to agree to a five-power conference on Korea and Indo-China, to take place in Geneva in the latter part of April. The upcoming meeting was widely thought of as Eden's conference, and the last thing he wanted right now was for Churchill to go off for talks that threatened to steal some of that precious limelight. Eden told the prime minister that he considered the message from Moscow an insulting way to put things to Britain's leader. The most Churchill would concede was that while he would not attempt to pursue the idea of bilateral talks before Geneva, he might well return to it down the line.

Eisenhower checked in two days later. The answer, again, was negative. But while Churchill had been awaiting his response, word had broken of a new US hydrogen bomb test. At the beginning of the month, a Japanese fishing boat had been deluged with radioactive ash though the vessel was some eighty miles outside the test zone at the time of the explosion. The news that twenty-two crew members had been burned and their cargo of tuna and shark poisoned provoked mass outrage in Britain. The British were so upset because they were so vulnerable to a nuclear attack. Eisenhower inadvertently made things worse when he disclosed at a press conference that the magnitude of the hydrogen blast had far surpassed scientists' expectations. The president's remarks, delivered in an incongruously by-golly tone, created the impression that the test had spun wildly and perilously out of human control.

Much had changed since Churchill in his message to Eisenhower expressed astonishment that the Cole disclosures had attracted so little public comment. Britons were now in full outcry and Churchill perceived an opportunity to cast himself as their voice. As he had done in May 1953, he planned to counter Eisenhower's intransigence by appealing directly to the people about the urgency of a summit meeting. This time, the occasion would be a House of Commons debate on the H-bomb on April 5. On that day, Churchill meant to regain the ground he had lost in Bermuda.

The opposition continued to support Churchill's Soviet initiative, but they were not eager to let him make political hay of the H-bomb debate. For days on end the Labour press upbraided Churchill for having allowed Britain to become a powerless, voiceless satellite of the US. Labour MPs amplified the charges on television and in the House of Commons. Why had Churchill acquiesced to the H-bomb experiments? Why had he taken no action to halt them? Why, again and again, did Britain's great ally act unilaterally? Why did Churchill not talk "straighter and harder" to the Americans? According to his attackers, the answer to each of these questions was that Churchill was too old and impotent to carry on.

"His battles are past," the mass-circulation *Daily Mirror* editorialized. "This is the Giant in Decay."

The Giant was not amused. At the end of the Second World War, Clementine Churchill, along with certain of Winston's friends, had been known to wince at the national hero's willingness, even eagerness, to shape-shift into a vituperative, if ineffectual partisan battler. Some Tories still blamed the Conservative rout of 1945 on the violence of Churchill's invective during the general election campaign, which had contrasted unfavorably with Attlee's "more statesmanlike" demeanor. In 1954, did Labour calculatedly goad the prime minister to self-destruct, as some opposition members later claimed to have done? However it happened, the angry apostle of peace arrived in the Commons on April 5 with an ill-considered speech that lurched between calling for three-power talks on the H-bomb—his principal purpose—and dropping a super-bomb or two of his own on the opposition.

Attlee spoke first, and wrong-footed Churchill with a grave, impeccably patriotic speech that was widely hailed as the most statesmanlike of his career. In contrast to the scurrilities of recent days, Attlee served up no criticism of Churchill. This afternoon at least, he had only praise for the prime minister, whose grasp of history and experience of world problems he painted as unsurpassed. Emphasizing that the international situation was suddenly too dangerous for politics as usual, Attlee insisted that he offered his party's resolution "in no party spirit" and that he sought "no party advantage." The opposition called for an immediate initiative by Her Majesty's Government to put Churchill together with Eisenhower and Malenkov.

Was that not exactly what Churchill wanted? As he was the first to admit, there was almost nothing in the opposition leader's speech with which he could find fault. When Attlee sat down, it was widely expected that the prime minister would follow with an even better speech, though in the same noble, nonpartisan vein. Many listeners expected that he would deliver another "epoch-making" performance along the lines of May 11.

Instead, hardly had Churchill begun when he launched into a slashing partisan denunciation of the Attlee regime for having abandoned Britain's veto over the use of the bomb. The loss of the veto had been eating away at him since he sat helplessly opposite Eisenhower that terrible day in Bermuda, when the president spoke of unleashing atomic weapons. Today, Churchill asked why Labour blamed him for a lack of influence, when it was they who had fatally weakened Britain's hand. As it became apparent that he was in full attack mode, the House went wild.

The opposition pelted Churchill with cries of "Resign!" and "Retire!" They accused him of acting disgracefully, of dragging the debate into the gutter, and of sacrificing the interests of humanity to make a cheap party point. The cacophony seemed to stagger Churchill physically. Unable to hear what some of the members were shouting at him, he looked helpless and confused.

Overcome with frustration, Churchill stretched out both hands and beseeched the House, "I have a right to be heard!" The opposition persisted in hooting him down.

Macmillan guessed that a few years previously, Churchill, though never the nimblest of debaters, would still have been flexible enough to readjust himself to the situation and depart from his script. Today, he dutifully, doggedly hewed to the prepared text. White-faced, he swayed slightly as he soldiered on in a tired, tremulous voice. Most Tories sat out their leader's martyrdom in grim silence. Woolton, concerned about the strain on Churchill in his present state of health, was relieved to spot Moran in the visitors' gallery. The doctor worried that the ordeal might trigger another stroke. Boothby drew a good deal of attention by flamboyantly rising from his seat below the gangway, turning his back on the prime minister in what many witnesses interpreted as a display of naked disdain, and marching out of the chamber. Boothby later claimed to have undergone periodontal surgery earlier in the day and to have had to leave suddenly on account of the pain.

People of all political stripes agreed that Churchill had made a fatal mistake in exposing his debility to the House and the public. He had intended to smite his enemies, but the only person he managed to damage with his cringe-inducing performance was himself. He had hoped to instill confidence, but he had undermined it instead. He had previously conceded that he could not go on after his stroke if he failed to exhibit his usual mastery of the House. Lord

Layton, a leading member of the Liberal Party, predicted that the day's brutal spectacle would hasten the prime minister's resignation. Had Churchill fought his way back from the brink of death only to commit political suicide?

The H-bomb debate raged on for several hours more. Eden, set to wind up for the Tories, conferred with fellow Conservatives in his room at the House. In recent weeks, "poor Anthony" (as colleagues referred to him) had grown so exasperated with Churchill that he seemed at times to be on the verge of another physical collapse. Two ministers recommended that he make no effort to rescue Churchill. Churchill had brought himself down; why not just let him stay down?

Under the circumstances, Eden preferred to be kind. That evening, he took the high road that Churchill had so disastrously spurned earlier. He argued that the prime minister had had every reason to be upset about the vicious attacks on him in the Labour press. Eden's remarks may have done little to help Churchill, but they immediately and immeasurably enhanced the heir apparent's standing in both the party and the House. Where Churchill had been rabid, Eden had seemed measured. Where Churchill seemed scattered and confused, Eden came off as reassuringly controlled.

The morning after Eden spoke up for Churchill in public, he moved against him behind the scenes. Having magnificently defended his chief's April 5 performance, he now offered it as evidence that Churchill must be made to go. Previously, Eden had been careful to distance himself from the machinations. It was a measure of how much the prime minister's stock had fallen that Eden was now actively driving the efforts to remove him. In the course of the day, he checked in with Butler, Salisbury, Macmillan, and others. That evening, he, Buchan-Hepburn, and Brook (who as recently as Bermuda had maintained that it was still best that Churchill stay on) agreed that Churchill must leave office no later than Whitsun, June 6. Eden and his colleagues also concluded that so long as Churchill remained in power he must not under any circumstances be allowed to attend a summit meeting.

As if he were living in a parallel universe, Churchill viewed April 5 very differently from most people. Where others heard taps, Churchill heard reveille. Instead of focusing on his mortifying performance and on the personally devastating newspaper coverage, Churchill riveted his attention on an indisputable fact: in the end, the opposition motion calling for the prime minister to press for a Churchill-Malenkov-Eisenhower meeting had carried unanimously. As far as Churchill was concerned, that gave him a "warrant" to carry on. Ironically, Eden and other leading Conservatives who saw the debacle as proof that Churchill was not fit to meet the Soviets had just voted along with the rest of the House instructing him to seek a summit meeting. Churchill thanked Eden for his speech (which, he took note, had endorsed the motion) and suggested that he was more disposed on account of it to regard Eden as fit to lead. The timetable Churchill had in mind for the succession, however, did not coincide with Eden's plans. After April 5, Churchill took a new line: how could he even consider handing over until he had accomplished what the House had demanded of him?

The day after the H-bomb debate, Eisenhower asked Churchill to join him in threatening the Chinese in Indo-China. "I know of no man who has grasped more nettles than you," the president wrote to him. "If we grasp this one together I believe that we will enormously increase our chances of bringing the Chinese to believe that their interests lie in the direction of a discreet disengagement." Should China refuse to discontinue support of the Vietminh rebels, Eisenhower wanted Britain to be prepared to participate in military intervention. Coming on top of Eisenhower's talk at Bermuda of unleashing missiles on Chinese bases, this new proposal horrified Churchill. At the same time, he was not anxious to displease the Americans just when he was gearing up to come back to them with a new summit pitch.

Eden too was appalled, both because he also vehemently disapproved of the American proposal and because he feared its potential impact on his own immediate plans. He had been hoping that a productive Geneva Conference would set up his homecoming as Churchill's successor nicely. The last thing he wanted was for China and the Soviet Union to pull out at the last minute. Both Churchill and Eden were conspicuously antsy and conflicted when Eisenhower's request was discussed at the April 7 Cabinet. It therefore fell to Salisbury to voice the most forthright objections. Salisbury demanded to know exactly what Eisenhower meant when he spoke of "air action" against China. Was he referring to atomic bombs?

As Churchill stealthily crafted a new campaign to sell Eisenhower on a bilateral meeting, he was careful to take the president's psychology into account. In the past, when Churchill had spoken of a solo mission to see the Soviets, he had cast the suggestion in the form of a threat, as the alternative strategy it would be his duty to pursue if the president refused to accompany him. This time around, he repackaged it as a favor. If Churchill struck out with Malenkov, the onus would be solely on Churchill. If, however, he managed to secure the token of good faith Eisenhower had demanded, the president would have the political cover he needed. He would be able to go to the summit without fear of criticism that he had allowed himself to be conned by the Soviet leadership.

There was another significant difference between Churchill's new proposal and past appeals. To date, Churchill had been seeking Eisenhower's immediate commitment to top-level talks. Here, he split his request in two. First he planned to ask Eisenhower to give him the go-ahead to see the Soviets alone. If, however, Churchill managed to secure an Austrian treaty (one of the proofs of goodwill Eisenhower had suggested in his April 1953 speech), only then would he seek a commitment to three-power talks.

Churchill knew that given his own age and health, this might well be his last chance to win over the president. It was likely now or never, so he was determined to argue the point in person. On April 22, without advising anyone, he wrote Eisenhower to suggest that he come to Washington in May. Churchill said only that he would "very much like to have some talks." About what specifically he did not say. "I should keep the plan secret till the last moment.

Do you like the idea?" For the past few weeks the White House had continued to militate for British support in Indo-China, so Eisenhower naturally assumed that this was what the prime minister wished to discuss. Churchill for his part was happy to have Eisenhower believe so. At length, Eisenhower suggested he come in June. When Churchill wrote to inform Eden of the trip, without of course spelling out what he hoped to accomplish in Washington, it became clear that there was no longer any possibility that he would hand over before Whitsun as the heir apparent and his supporters had been hoping.

Churchill told the Cabinet about his trip on June 5. Addressing the ministers, he was as vague and evasive about his purposes as he had been with Eisenhower. He indicated that he wished to talk to the president about atomics and Egypt. About his new summit proposal, however, he uttered not a syllable. He noted that he had originally planned to have Eden accompany him to Washington, but as the Geneva Conference was dragging on he now proposed to go alone.

Salisbury would not hear of it. In Cabinet discussions to date, Salisbury had been a fierce advocate of doing everything possible to discourage the US from "precipitate action," whether in Indo-China or other hot spots. Salisbury's concerns had resonated strongly with Cabinet colleagues. He regarded Eisenhower as a "prize noodle" and he worried desperately that even without British support the US might bomb China on the rationale that the international situation threatened to become much more dangerous when the Soviet Union was further along in the nuclear arms race. Salisbury maintained that, on the contrary, the danger of war would be less when there was "a strong deterrent to all." In the meantime, he argued, the British must make it their mission during the next two or three years to stay close to the Americans. At the June 5 Cabinet, Salisbury pointed out that Indo-China was sure to be spoken of while Churchill was in Washington. Salisbury wanted Eden to be there to counter any arguments Dulles might make in favor of military intervention.

In the ensuing discussion round the Cabinet table, there was a subtext of concern that it really would be best if Eden (who was present) went to Washington in his longtime capacity as Churchill's restraining hand. No one said as much in front of the old man, but the feeling in the room was palpable. Churchill raised no objections, and at length it was agreed that Eden would somehow arrange to go as well.

After the Cabinet, Churchill and Eden had a private talk about the succession. Churchill said he now intended to leave office at the end of July when Parliament rose for the summer recess. Eden protested that that would not be an ideal time to form a new government, but Churchill was, or pretended to be, oblivious to his objections. Later, Churchill told his wife how well the talk had gone. He insisted that he and his successor had reached "a very perfect understanding." Eden certainly did not see it that way. On his return to Geneva, he wrote to complain further. He asked Churchill to step aside as soon as he came back from Washington, preferably before the end of June. The request so offended Churchill that he withdrew his offer to go in July. Suddenly, he refused to commit to any date at all. The most he would say was that he hoped he would not have to remain in power beyond the autumn. He informed Eden that he must "see what emerges from our talks in Washington and how they affect the various schemes I have in mind."

On June 25, 1954, when Churchill emerged from a black Lincoln convertible at the North Portico of the White House escorted by Vice President Richard Nixon, neither the Americans nor the Cabinet in London had a firm grasp of his intentions. In the run-up to the Washington meetings, Sir Roger Makins had wittily observed that the Americans tended to be "terrified" of Churchill's visits and that they liked to take protective actions, "like squids or skunks," in advance. To judge by Churchill's fragile appearance, however, there seemed to be nothing very much to guard against. In contrast to the beaming, effusively smiling Eisenhower, who trotted down the marble steps to welcome him, Churchill ascended with the gingerliness of great age. His head was bent and his eyes, with their pale lashes, were downcast lest he stumble. At moments, he resembled a marionette whose cords had been cut. As Eisenhower hovered protectively at his side, Mamie Eisenhower, standing at the top of the stairs, leaned forward, grasped Churchill's hand, and helped guide him the rest of the way up.

Offered a choice of accommodations in the family quarters, Churchill asked for the Queen's Suite, where he had stayed during the Roosevelt presidency. When he was shown to his rooms, which had rose-colored walls and a canopied four-poster bed, his hosts assumed he would need to rest before the business of the conference began. Eden and Dulles closeted themselves in the presidential study on the second floor of the White House to discuss Germany. They too assumed that it would be at least an hour before the prime minister was ready.

Churchill surprised everyone. Given his admittedly waning strength, it was far from ideal that, as matters stood, he might not have an opportunity to make his summit case to Eisenhower until the end of seventy-two hours of tightly scheduled meetings on Indo-China, Europe, and a raft of other topics. Thus, as soon as he reached his rooms, he changed into a fresh shirt and fawn-colored tropical suit and waddled off in search of the president. When the leaders found themselves alone in the Oval Office for three-quarters of an hour, Churchill briefly outlined his new summit proposal. This time, he found that he did not have to do very much. He was elated, relieved, and not a little flabbergasted when Eisenhower's answer—immediately—was yes. Eisenhower accepted that Churchill would see Malenkov first. If Churchill managed to wrest a token of good faith from the Soviets, Eisenhower would join in later.

Why the abrupt pivot? It was not really anything Churchill said or did. Rather, it was what Eisenhower expected to get from him that seems to have made the difference. While Churchill was at the White House, Eisenhower hoped—vainly, it turned out—to convince him to sign off on a unified Anglo-American position on Indo-China. Congressional support for intervention was contingent on British involvement. What harm could there be in putting his guest in a good humor by agreeing in principle to a summit meeting that was never going to happen on Churchill's watch anyway?

By the look of him, Churchill was unlikely to be in office long enough to arrange two-, let alone three-power talks. Even if he managed to hang on, Eden and the Cabinet were sure to block him before things went too far. Besides, Churchill could not take any further action until he had checked in again with the White House, could he?

Churchill emerged from the encounter in roaring spirits. The official sessions had not even begun, but as far as he was concerned he already had what he had come to Washington for. Eisenhower, by contrast, was not so pleased. Hardly had he assented to Churchill's proposal when he began to have second thoughts. When, after lunch, he and Churchill met with the others he struggled perceptibly to slow things down. He asked Churchill to put his proposal in writing and he suggested that it would be best to reach out to the Soviets through normal diplomatic channels.

Churchill sensed what Eisenhower was trying to do. But like a dog with a bone, he was not about to give up what he had won earlier; he would not let Eisenhower turn that yes into a no. Churchill nimbly sidestepped the president's many cavils and caveats. He offered only enough information so that later no one could claim that he had failed to spell out his intentions. He mentioned no dates other than to assure the Americans that he would do nothing until he was back in Britain.

Finally, after repeated efforts to regain control of the situation, Eisenhower sent in the reinforcements. The prime minister had mentioned a "reconnaissance in force" and Dulles had privately told Eisenhower that he found the idea alarming.

"You are to be a go-between," said Dulles with undisguised distaste when he sat down with Churchill on the third day of the visit.

"No," Churchill replied cheerily. "For I know on which side I stand. I will be a reconnoitering patrol."

Dulles expressed doubt that Churchill could get an Austrian treaty. The foreign ministers had already tried that with Molotov. He warned that were Churchill to fail as well, an impression might be created in the world that the only alternative was war. He predicted that a solo mission would not play well in America and that the White House might be compelled to announce that Churchill did not speak for the US. Though—crucially, in Churchill's view—Dulles never asked him in so many words not to contact the Soviets, the secretary of state urged that the matter be "very carefully weighed before any positive decision was made."

Afterward, Dulles reported the conversation to Eden when the foreign secretary visited him at home in Georgetown. Relations were icy between Dulles and his British counterpart. They had collided often and acrimoniously in Geneva, where Eden had been at pains to convey that British foreign policy would be set in London, not Washington. Though Dulles was rooting for Butler to succeed Churchill, he knew Eden to be no advocate of top-level talks and therefore a useful ally. Previously, Eden had looked to Eisenhower to block Churchill's path to the summit. Now, the Americans were looking to Eden.

The long-suffering protégé was indeed eager for a heart-to-heart with Churchill after they left Washington. But it was the succession, not the summit, he burned to discuss. Had Churchill not said that that conversation would have to wait until he had finished with Eisenhower? Eden, who had been scheduled to fly home, decided to join Churchill on his ship in the hope of extracting a firm date for the handover.

Where Eden looked upon the Washington visit as an end, Churchill, characteristically, saw it as only a beginning. In contrast to his mood after Bermuda, he left the US a happy man. Churchill knew full well that Eisenhower was far from pleased and that there might be trouble down the line, but by his lights he had the president's blessing nonetheless.

At least, he could claim that he did. Having got what he needed at the White House, Churchill panted to be in touch with the

Kremlin. As far as he was concerned, he did not have time to waste on the five-day sea voyage his doctor had strongly recommended for his health. Besides, he was in no mood for another confrontation with Eden. At the last minute, Churchill thought he might duck the heir apparent and fly home instead.

XVIII

AN OBSTINATE PIG

Aboard the Queen Elizabeth, *July 1954*

In a loud voice that carried through the small, hot, crowded grill room on the sun deck of the *Queen Elizabeth*, Churchill protested that he did not know why he had allowed himself to be persuaded to travel by sea. It was lunchtime, July 1, 1954. The prime minister's party had boarded in New York the night before, and Churchill grumbled that he would have been at Chartwell already had he flown.

Angrily, he complained that the restaurant had no fan, that the windows were sealed, and that the sun was beating on him through the glass. Suddenly, to the embarrassment of his companions, he fled to another table, and then another. But nothing seemed to comfort him.

Seated at the original table with Soames and Moran, Colville suggested that the real reason Churchill was so on edge was that he knew matters were about to come to a head with Eden. Indeed, that night a "bashful" Eden sought Colville's advice about the best time to speak to Churchill.

Meanwhile, Churchill had a chance to completely rethink his attitude toward the sea voyage. Instead of regarding it as a waste of precious time, he came to view it as an opportunity. And instead of being aggravated and offended by Eden's impatience, he found a way to use it to his advantage.

Churchill faced a huge problem at home. He had not told the Cabinet beforehand about his new summit pitch, and he expected they would raise objections and cause delays. So, though he had assured the Americans that he would wait until he reached London, he decided to contact the Soviets from the boat. Cabinet ministers would no doubt be incensed when they learned what he had done, but by then it would be too late to stop him. The scheme had the further advantage of boxing in the Cabinet in the event of a favorable response from Moscow. At that point, ministers would find it difficult to block Churchill without endangering themselves politically.

On the morning of July 2, when Eden finally worked up the courage to approach him, Churchill was unusually welcoming. He offered to retire in September—if, that is, Eden made no objection to his contacting the Soviets to float the idea of a meeting. When Eden agreed to the horse trade, Churchill closeted himself with Colville to dictate a telegram to Molotov. He proposed two-power talks that could lead to a reunion of the Big Three "where much might be settled." How, he wanted to know, would Moscow feel about such a plan? Churchill was proceeding on the assumption that after three years of stall tactics the French were finally about to ratify the EDC by the end of the summer. With the German problem resolved, he could see Malenkov in late August or early September—in time (theoretically anyway) to deliver on his part of the bargain with Eden. It was a tight schedule, to be sure, but it just might work.

Tempted with a definite date, the restraining hand had been unable to restrain himself. Yet hardly had Eden made his deal with the devil when second thoughts consumed him. After Churchill, Eden, and the others in their traveling party had a high-spirited lunch in the prime minister's dining room, Eden tried to claw back what he had agreed to earlier without jeopardizing the September date. Eden urged Churchill to wait until he had had a chance to check in with the Cabinet, in keeping with constitutional proprieties. He offered to personally deliver Churchill's message to Molotov when the foreign ministers reconvened in Geneva.

Churchill refused to consider it. He insisted that his message was not an official proposal and that he was only asking if the Soviets would care to have a visit. If the answer was no, he need not trouble the Cabinet. If it was yes, and the Cabinet disapproved he could easily get out of the whole thing by notifying Molotov of his colleagues' point of view.

The crossfire continued all afternoon and evening. Eden persisted in recommending delay. Churchill demanded to telegraph at once. Eden warned that if Churchill went forward it would be contrary to his strong advice. When Colville and Soames sided with Eden, Churchill angrily threatened to throw both his private secretary and his son-in-law out the porthole.

For all the bluster on Churchill's part, there was a good deal of quiet calculation. At length, he offered to show the telegram to Butler and to invite his comments—if, that is, Eden would allow him to say that he agreed with it "in principle" (which, they both knew, he did not). By his acquiescence, Eden permitted Churchill to make him complicit in the decision to act before they reached London. Churchill had fashioned a piece of evidence that he had proceeded not just with Eden's knowledge, but with his support.

Churchill ran a considerable risk, however, that Butler might decide on his own to summon a Cabinet. That was why Churchill worded his covering message as he did: "I propose to send this personal and private telegram, with which Anthony agrees in principle, to Molotov. I hope you will like it. The matter is urgent." The telegram went off to London at half past ten that night. Churchill hoped that Butler would get back to him immediately, while Eden hoped that he would share the document with Cabinet colleagues and that they would move in unison to stop Churchill before it was too late. Eden did not, however, press Churchill to direct Butler in so many words to show the telegram to anyone else.

Everything depended on Butler, but by the following afternoon he had yet to respond. Eden's and Churchill's behavior in the face of his thundering silence was typical of both men's natures. Eden pouted about how badly Churchill had treated him the previous day, but otherwise waited passively. Churchill grumbled that he had allowed himself to be trapped into querying Butler, but he also took action to make things go his way. The longer the delay, the likelier it seemed that Butler had summoned a Cabinet. On the chance that Butler had yet to do anything irrevocable, Churchill fired off a deviously worded telegram at a quarter to three that afternoon: "Presume my message to the Bear has gone on. It in no way commits the Cabinet to making an official proposal. Time is important."

For the moment anyway, time was on Churchill's side. This was a Saturday and Butler was in Norfolk for the weekend. Churchill's first telegram had arrived in London shortly after midnight, but it had had to be sent to the country by dispatch rider and did not reach the chancellor until nearly 5 p.m. on Saturday. That evening, Butler conferred with the Foreign Office. On their advice, he was about to urge the prime minister to wait until he had had a chance to talk to the Cabinet when Churchill's misleading second telegram arrived. Told that Churchill assumed he had already sent on the Molotov telegram, Butler fretted that he had somehow misunderstood everything. Without delay, he sheepishly sent Churchill some suggestions for minor edits, along with the assurance that he would transmit the message to Moscow as soon as he heard back from him.

Churchill had brought it off. Not only had he managed to circumvent the Cabinet, he had also added the chancellor to the list of those who had had a hand in the affair. On Butler's instructions, the British ambassador in Moscow, Sir William Hayter, went to the Kremlin on Sunday afternoon to deliver Churchill's missive.

Two days later, the *Queen Elizabeth* docked in Southampton, where Churchill and Eden boarded a special train to London. Butler was waiting at Waterloo Station at seven that evening, and he accompanied them to Number Ten. There, Churchill hardly had a chance to catch his breath when the Soviet ambassador, Jacob Malik, arrived with a message from Molotov. Churchill had not expected to have his reply so quickly. Molotov wrote, "Your idea about a friendly Meeting between you and Premier G. M. Malenkov as well as the considerations expressed by you regarding the aims of such a Meeting, have met with sympathetic acknowledgement in Moscow."

Molotov's answer sent Churchill into overdrive. Suddenly, he had what he believed he needed from both the Americans and the Soviets. Ironically, the main obstacle that faced Churchill now was his own Cabinet.

The prime minister had already entangled Eden and Butler in his web. Immediately, he moved to draw in Salisbury and Macmillan as well before he faced the full Cabinet the next morning. In the past, both men, Salisbury in particular, had vehemently opposed any talks that did not include the Americans. It was after 10 p.m. when Churchill summoned Salisbury and Macmillan to the Cabinet Room. Ostensibly, he wanted them to consult, along with Eden and Butler, on the delicate task of informing Eisenhower that he had already contacted Molotov. But since when did Churchill need help crafting wily telegrams?

Salisbury arrived first. Macmillan, who had just been getting into bed when the phone rang, entered a bit later. Eden and Butler, both looking the worse for wear, were also at the table. The file the prime minister passed around, comprising the shipboard messages to Butler and Molotov, did not display either Churchill or Eden in an especially flattering light. Clearly, Churchill had flouted constitutional proprieties by acting on a major foreign policy initiative he had yet to discuss with the Cabinet. Clearly, he had been far from straightforward in his arrangements with Butler. And clearly, Eden had countenanced Churchill's actions, whether out of weakness, ambition, or both. After Salisbury examined the documents, he was visibly distressed. But it was only when Churchill spoke of the need to explain himself to Eisenhower that the full dimensions of what the prime minister had done began to sink in.

It was not just the Cabinet whom Churchill had put in an awkward position by presenting his correspondence with Molotov as a done deal. That in itself was bad enough. Even worse was the fact that Eisenhower had as yet no inkling that Churchill had made an overture to Moscow though he had been the president's guest only a few days before. While Churchill believed that he had acted in the interest of peace, Salisbury judged that his behavior at sea had made the world more rather than less dangerous. In the interest of minimizing the damage to the Anglo-American alliance, Salisbury stifled his indignation and joined forces with Churchill in an effort to finesse Eisenhower.

At length, the ministers approved a message to Eisenhower that was as guileful as those Churchill had written to Butler from the boat: "In the light of our talks and after careful thought I thought it right to send an exploratory message to Molotov to feel the ground about the possibility of a two power meeting. This of course committed nobody except myself." Churchill spoke as if his actions flowed directly from the Washington talks and as if he had complied fully with Dulles's request that he consider carefully before he took a decision. He pretended that he had not said that he would wait until he returned to Britain, and he prompted Eisenhower to comment, not on what he, Churchill, had done, but rather on Molotov's reply. What, he asked, did Eisenhower make of the message from Moscow?

Churchill had moved slyly and swiftly to bind the president to him, to make Eisenhower a partner rather than an opponent. Whether the stratagem would work remained to be seen. At the same time, as Macmillan perceived, Churchill had acted along similar lines with Salisbury and himself.

The ministers left Number Ten after midnight. Salisbury and Macmillan accompanied Eden to the Foreign Office to hear his side of the story. It turned out that they were not the only ones who were disappointed in Eden. Awaiting him was a stack of messages from Foreign Office colleagues who were appalled by Churchill's actions on the boat and troubled by Eden's failure to restrain him. Eden alternated between insisting that "no power on earth" could have stopped Churchill and blaming Butler for his failure to go to the Cabinet when he had the chance.

Macmillan wondered why Eden had not simply threatened to resign, the final option when one cannot live with a prime minister's decisions. But then, had Churchill called his bluff, Eden would have left office just when it seemed as if his accession was—really, this time—about to take place.

Churchill's telegram went off to Washington at a quarter to three in the morning. At half past eleven, he was back in the Cabinet Room. Of the fifteen ministers seated around the table, only those who had attended the previous night's session at Number Ten knew about the "bombshell" he was about to drop. The others were still "blithely unconscious" of what awaited them. This was the first time most had seen Churchill since his journey, and the room echoed with congratulations on a job presumably well done. Those few in the know made no effort to disillusion their colleagues; they would find out soon enough. For an hour and ten minutes, Churchill meandered through various less incendiary topics before he came to the fifth and penultimate item on the agenda—the Washington talks. As he began to speak, his nervousness was perceptible. Before long, the reason for his unease became clear.

He opened by saying that while he was not ready to give a full account of his trip he "must, however, inform the Cabinet of a project for a meeting with Russia." As he offered ministers their first glimpse of his retooled Soviet initiative, he emphasized that Eisenhower had admitted his right to go and that Dulles had "tacitly" accepted the possibility that he would go. Churchill went on to disclose that he had contacted Molotov from the ship and that he had already had an answer.

His audience reacted to these revelations with looks of "blank surprise." Crookshank, who sat to one side of the prime minister, radiated "disgust." Churchill sought to obfuscate his failure to consult the full Cabinet beforehand by highlighting the various consultations he had undertaken—with the foreign secretary, the chancellor of the exchequer, the Lord President and the minister of housing. None of this was enough to dissipate the gathering storm.

When Churchill finished, he asked Eden to comment. Eden spoke briefly and anxiously. He said a few words about possible European reactions and he seconded Churchill's point that it would be best to wait to hear back from Eisenhower before the Cabinet took a decision about a reply to Molotov. But what, really, could Eden say after he had allowed Churchill to go forward on the boat? As at the time of his 1938 resignation speech, there was the uncomfortable appearance of Eden having been at pains to guard his claim to the succession.

Also as in 1938, it fell to Salisbury to speak the harsh words that Eden dared not utter. Clearly, Churchill's attempt to defang Salisbury had failed. Just when a parley at the summit seemed achingly within grasp, Salisbury put Churchill on notice that he was prepared to stop him. Acting as if he had no doubt that his was the opinion that counted in the room, Salisbury stated that he had reflected overnight and that his attitude would turn largely on the "temper" of Eisenhower's answer. If the answer were "critical or hostile," Salisbury would advise the Cabinet to vote against going forward with the Malenkov meeting because it would damage "the US/UK relationship." That relationship, Salisbury made it clear, overrode everything else. "Meanwhile, I reserve my attitude."

Macmillan began to speak, but Churchill cut him off to counter what Salisbury had just said. "The United States can't veto my visit," Churchill pointed out. "They accept that." The prime minister made no reference to what Salisbury had heard Dulles tell him the previous October—that, while the Americans accepted that the decision was wholly Churchill's to make, a solo mission would jeopardize the Anglo-American alliance.

After the Cabinet agreed to wait to hear from Eisenhower, Churchill turned to the last item on the agenda, the question of whether Britain ought to manufacture its own hydrogen bomb. He laid out the strategic argument that such a weapon would assure Britain's place at the table in all subsequent East-West negotiations. But he mischievously left it to Salisbury to disclose that the defense policy committee had already decided to build an H-bomb and to carry on with active preparations. In his capacity as Lord President and minister for the bomb, Salisbury had participated in the secret deliberations in June. The revelation that there had been no Cabinet consultation on this matter as well left several ministers reeling. They were upset on account less of the decision itself than of the improper way in which it had been made. When Crookshank angrily protested that such a momentous decision had been communicated to the full Cabinet in so cavalier a fashion, it became evident that Churchill was up to his old tricks. He had set the birds against one another. Crookshank stalked out, others followed, and the meeting broke up unceremoniously.

By this pretty maneuver, Churchill put Salisbury on notice that if the noble lord really meant to go up against him, he had better watch his back. Macmillan spotted a disaster in the making. If two such uncompromising characters went to war, the effect on both the government and the party could be disastrous. All this was coming at a most inconvenient time for Macmillan personally. Just when the post of foreign secretary in an Eden government was conceivably about to be his, he was distraught at the prospect that the government might fall.

Churchill, by contrast, seemed as though he had not a care in the world when he hosted a dinner party in honor of Pug Ismay that evening. He knew precisely what he had and had not said in Washington. He was sanguine that in the end Eisenhower would accept that it was not within his power to prevent Anglo-Soviet talks. He was also optimistic that Salisbury could be mollified and that the crisis would end with a whimper. After dinner, Macmillan, who was one of the guests, had a talk with Colville about (what else?) the Cabinet crisis. When Colville pointed out that Churchill intended to retire in September in any case, Macmillan warned that the Government and probably the Tory party would have broken up before that. Macmillan alternated between viewing Churchill's actions on the boat as evidence of almost demoniacal cunning on one hand and of senility on the other.

"He must go at once, on grounds of health, to avoid a disaster," Macmillan declared.

"You are very severe," Colville protested.

"As you know," said Macmillan, "I am devoted to Winston and admire him more than any man. But he is not fit. He cannot function. If there were a strong monarch, of great experience, he would be told so by the palace."

When Colville refused to ask Churchill to leave office before September, Macmillan proposed that he at least speak to him about the gravity of the crisis. "I beg of you to urge him not to try to ride this one off too easily. He must take it seriously and realize how deeply he has hurt us."

Meanwhile, Churchill's telegram had reached Washington. Far from sharing Churchill's delight in Molotov's friendly reply, Eisenhower regarded it as proof that Moscow aimed to drive a wedge between the Americans and the British. No one, not Eisenhower or Dulles, not Eden or other Cabinet ministers, had been able to stop Churchill. The one remaining hope seemed to be a higher power. Eisenhower and Dulles agreed that "only another stroke" would be capable of preventing Churchill from going to Malenkov.

Eisenhower complained to Makins that Churchill had confronted him with a tricky political problem. Anglo-Soviet talks were sure to play poorly with the American people. In anticipation of the inevitable questions from press and public when the news broke, the president wondered whether it might not be best if he simply "went fishing." Eisenhower ruefully acknowledged to Makins that in the end he "could not prevent the head of a friendly state from doing what he thought right."

Before the day was done, Eisenhower wrote to Churchill in a tone of annoyance: "You did not let any grass grow under your feet. When you left here, I had thought, obviously erroneously, that you were in an undecided mood about this matter, and that when you had cleared your own mind I would receive some notice if you were to put your program into action." Eisenhower conceded, however, that that was now "past history." Treating it as a foregone conclusion that Churchill would soon publicly announce his plans, Eisenhower asked only to be told the date so that he could prepare a statement of his own. He would probably say that while Churchill was in Washington the possibility of a Big Three meeting had been spoken of, that he had been unable to see how it could serve a useful purpose, and that he had indicated that if Churchill did undertake such a mission the plan would carry Eisenhower's hopes for the best but that it would not engage his responsibility.

Eisenhower's missive reached the Foreign Office after midnight. When he read it, Churchill was thrilled. He had endured enough of Eisenhower's emphatic no's in the past to recognize that this message was something else entirely. And he sensed that with a bit of fancy footwork he might be able to coax an even more useful answer out of the president. When he began a new telegram, "I hope you are not vexed with me for not submitting to you the text of my telegram to Molotov," he was prompting Eisenhower to say that of course his anxieties were baseless.

In this slippery vein Churchill continued, "I felt that as it was a private and personal enquiry which I had not brought officially before the Cabinet I had better bear the burden myself and not involve you in any way. I have made it clear to Molotov that you are in no way committed. I thought that would be agreeable to you, and that we could then consider the question in the light of the answer I got."

Did Churchill really expect Eisenhower to believe much of this? Probably not, but at least there was the veneer of a decent explanation and perhaps they could proceed from there. Now that a parley with the Soviets was in play, Churchill suddenly was all for full and frank consultation with the Americans. He promised not to seek an official decision from the Cabinet until he had heard again from Eisenhower. And he insisted that there would be no question of an announcement until their two governments had consulted together and agreed on what it would be best to say.

When Churchill faced the Cabinet on the morning of July 8, he read aloud the full text of Eisenhower's telegram. Then he read his own draft reply and suggested that they wait to take a decision until they had had Eisenhower's further comments. He voiced confidence that he could get a "better" answer from the president—that is, an answer that put to rest any fears about Anglo-American relations.

Churchill went on to paint himself as blameless in having contacted Molotov without Cabinet consultation. He argued that the telegram had been a personal message no different from those he had exchanged with foreign leaders in the past. He emphasized that he had long reserved the right to communicate personally with his counterparts abroad, and he denied that in the present instance he had felt any obligation to go to the Cabinet.

The truth of course was more complicated. A few days after this, he privately confessed to Colville that he had deliberately gone around the Cabinet. Offstage, Churchill disclosed his abiding belief that had he waited to consult them until after his ship returned, they would almost certainly have held him up with objections and delays. He suggested that the stakes had been so high and the possible benefits so crucial to human survival that he had been prepared to use any methods to secure a meeting with Malenkov. Throughout Churchill's career, his willingness to take ethically questionable shortcuts had incensed countless critics. If he thought he was doing something important, his attitude was let the rules be damned. He dismissed those who had a problem with this approach as nitpickers.

Salisbury had been wincing at Churchill's bad behavior since at least the 1930s. This morning, Salisbury weighed in that Churchill's telegram to Molotov had constituted an important act of foreign policy which involved the collective responsibility of the Cabinet. Taking Churchill at his word that he believed he had the "absolute right" to conduct such correspondence, Salisbury pointed out that it was the remedy of ministers to resign if they disagreed. He did not say that he or any other minister would resign, with all that that might mean for the viability of Churchill's postwar government—only that it was an option.

At first, Churchill ignored the fact that the threat of resignation was now on the table. (As far as he was concerned, the Cecils had long debased the currency of that threat by using it too often.) Resuming his self-defense, Churchill argued that what he had done on the ship was less a fully fledged act of foreign policy than an enquiry. Action, he conceded, would have required Cabinet consent. Enquiry, on the other hand, was quite permissible so long as he showed the telegram to Eden. Then the zinger: "He [Eden] could have insisted that it should come to the Cabinet."

How was Eden to respond? As Salisbury and other supporters well knew, he had insisted on no such thing. Today, Eden's feeble claim that he had had it "in mind" that the chancellor of the exchequer would show the telegram to the Cabinet prompted an indignant Butler to jump in. Creating a sideshow was a tactic that had worked for Churchill in the past, but Salisbury was determined not to let him get away with it this time.

"Only two days passed before the Prime Minister returned," he interjected, turning the heat back on Churchill. A few minutes before, Salisbury had made the pretense of taking him at his word. Now, he implied that whatever Churchill claimed about his thinking on the boat, he had really just been trying to ram his policy through before the Cabinet could object: "What was the urgency which precluded waiting for a proper consultation?"

Confronted with the truth, Churchill replied tartly, "I may have exaggerated the urgency of my hope for strengthening world peace." He went on to suggest that if the Cabinet broke on the constitutional issue, most Britons would side with him. After May 11, 1953, there could be little doubt about where the people stood on the question of a summit meeting, and he threatened to appeal directly to them. Let his adversaries cry that he had skirted the rules; Churchill was sure the public would not care. Swinton, Lyttelton and Stuart, the secretaries of state for Commonwealth relations, the colonies, and Scotland respectively, scrambled to cool things down. In spite of their efforts, Churchill went further. He hinted that Salisbury was not the only one who might soon bolt.

"If I found myself at variance with colleagues," Churchill warned, "I could resign." Like Salisbury, Churchill did not say he would go—only that his resignation was on the table, with all that that might mean for his colleagues' political futures.

At a time when those colleagues had been praying that Churchill would agree to retire, he suddenly recast the possibility of resignation as a threat. Yes, they all still wanted him to go, but not this way. If Churchill resigned on this issue he would torpedo the Conservatives' chances in the next general election by making them look like enemies of the peace. Churchill had been known to abandon his party before; why not now?

At this point, Churchill did not think he would have to do anything so drastic. He calculated that with a little help from Washington he could yet defuse the crisis. On the morning of Friday, July 9, when he received Eisenhower's second reply ("Of course I am not vexed"), he moved at once to put a few more key words into the president's mouth. Churchill opened his third message of the series, "I am very much relieved by your kind telegram which reassures me that no serious differences will arise between our two governments on account of Russian excursion or 'solitary pilgrimage' by me." He gilded the pill by promising not to agree to meet Malenkov in Moscow, and by reiterating his plan to seek a gesture of Soviet intentions in the form of an Austrian treaty. If Churchill elicited the answer from Eisenhower that he hoped for, what choice would Salisbury have but to back off?

At half past twelve that afternoon, the Cabinet reconvened, this time in Churchill's room at the House of Commons. Churchill read aloud Eisenhower's second telegram as well as his own draft reply.

Salisbury cut in, "I am unalterably opposed to a meeting with Russia without the United States."

Previously, Salisbury had indicated only that he would base his decision on Eisenhower's response. Now that he had seen how effectively Churchill was playing Eisenhower, he announced that he would oppose bilateral talks no matter how good the answer Churchill managed to extract. Directly addressing Churchill,

Salisbury went on, "I told you last year why I was opposed to this." (He meant the letter he had written to him after the Derby.)

Salisbury continued, "Other members of the Cabinet may think it is right to go forward. But, if they so decided, I should have to resign." As Salisbury said this, Churchill turned "dead white" one moment and "puce" the next. After Salisbury finished speaking, there was silence. Salisbury had just moved into open and direct conflict with Churchill over the issue dearest to the prime minister's heart. He had in effect asked their colleagues to choose between Churchill and himself.

Churchill took a few seconds to regain his composure. "I should greatly regret a severance," he said at last. "But I hope our private friendship would survive."

After the Cabinet, an anxious Eden asked Macmillan to come to him at the Foreign Office that afternoon. As the two soon agreed, much had altered as a consequence of the explosive morning session. The time had been when Salisbury plotted and planned to make Eden prime minister. Now Salisbury was poised to bring down the government on the very eve of Eden's accession. The time had been when Salisbury worried that Churchill had embarked on a dangerous game of chicken. Now Salisbury had begun to play a similar game himself. The time had been when Eden complained bitterly of Macmillan (of his rapacity for high office, his toadying, and other sins) to Salisbury. Now it was Salisbury whose actions horrified him, and Macmillan in whom he confided.

As ever, Salisbury presented himself as calmly willing to sacrifice everything for his principles. But was he really? If the government fell, he would still be Marquess of Salisbury, a figure in both the party and the nation. Men like Eden and Macmillan stood to lose much more. Under the circumstances, their priority, as they saw it, was to keep the government intact until Eden was safely at its head. Macmillan went to Hatfield House that evening to beg Salisbury on behalf of Eden and himself to do nothing rash. Eden was due to return to Geneva the following week and the last thing he wanted was for Salisbury to provoke a political crisis in his absence. Initially, Salisbury seemed willing to hold off until Eden came home. In the course of the weekend, he changed his mind.

Salisbury decided that if he were to avoid becoming complicit in policies he did not support, he must bail out sooner rather than later. Contemplating the speech he would deliver in the House of Lords to explain his resignation, Salisbury thought he would cite the telegram to Molotov and the May 11 speech as examples of Churchill's tendency to make major decisions without properly consulting his colleagues.

On the morning of Tuesday, July 13, Churchill received a third message from Washington. Eisenhower wrote: "You, of course, know that never for one moment would this create any difference between two Governments which are headed by you and me . . ." He went on to say that he could not of course undertake to deliver US public opinion, but he pledged to do his best to "minimize" the American people's unfavorable reaction to Anglo-Soviet talks. On reading this, Churchill was ready to do a victory lap round the Cabinet table. Later that morning, he read the telegram aloud to the Cabinet and distributed copies. As he had hoped, most ministers seemed relieved. The predictable exception was Salisbury, who appeared "glum and worried." At this point, it certainly looked as if Salisbury was not going to be able to carry the Cabinet. Churchill calmly announced that he would wait to call for a decision until the Geneva Conference wound up in about ten days. Though he planned to go forward in any case, his proposal to Molotov would be cast differently depending on the outcome at Geneva.

Following the Cabinet, Churchill roared that he did not "give a damn" if Salisbury resigned. Now that he had his trump card in the form of Eisenhower's telegram, he was sure that no matter what Salisbury did, the rest of the Cabinet would stay put.

Eden, Macmillan, Swinton, and other colleagues did not share the prime minister's optimism. They expected that Salisbury's defection would be followed by Crookshank's and then others. In the days that followed, various failed efforts to persuade either Churchill or Salisbury to back down culminated in a visit by Macmillan on July 16 to Clementine Churchill at Number Ten. It was no great secret that she longed for her husband to retire. Macmillan spoke to her of Salisbury's imminent resignation and of the likely consequences. He reported the feeling in the Cabinet that much as they all loved Churchill it would be best for both the party and the country if he handed over immediately.

Would she talk to him?

Though Clementine Churchill had never warmed to Macmillan, she was cordial and composed during the visit. When he left, however, she called Colville and angrily recounted all that Macmillan had said. Her indignation puzzled Colville. He asked whether she did not herself feel that Churchill ought to step down.

"Yes, I do indeed," she replied, "but I don't wish to be told that by Mr. Harold Macmillan."

Clementine Churchill tended to be fearless with her husband. She was known to scold him thoroughly when it suited her, and the old couple had many stormy scenes. Yet on the present occasion she preferred not to be alone with him when she conveyed his colleagues' views. Colville joined the Churchills for lunch, but the confrontation did not go well. Colville thought she would have been wise to lead off by pointing out that ministers were anxious to avert Salisbury's resignation. Instead, her comment that the Cabinet was angry with Churchill for mishandling the situation caused him to "snap back" at her. At length, Churchill turned to his private secretary and asked him to notify Macmillan that he would be glad if he would come and tell him in his own words what he thought—"rather than tell my wife."

When Macmillan reappeared, Churchill was in no mood to listen. He had already heard the particulars at lunch. Instead, he delivered an angry, rambling, repetitious defense of his actions on the boat.

As far as the Cabinet crisis was concerned, Churchill boasted that he held all the cards. He warned that neither the Conservative rank and file nor the country at large would tolerate a palace revolt by disaffected ministers. He pointed out that his personal popularity transcended party lines, and he promised that if Salisbury and others resigned he would simply form a new and more powerful government. Possibly, he would form a coalition government with the opposition, who, whatever their differences with him in other areas, wholeheartedly supported his Soviet initiative. According to Churchill, Salisbury's preoccupation with constitutional niceties would stir the public not at all. The only issue that would resonate was whether meeting Malenkov was a good idea.

When Churchill gloated that Salisbury's resignation would be as foolish as Lord Randolph Churchill's had been, he was referring to some highly charged family history. In 1886, Winston's father, who had been spoken of as a possible successor to Lord Salisbury as prime minister, tendered his resignation as chancellor of the exchequer. He claimed to be driven by differences on matters of high policy, but the Cecils insisted that his resignation had had nothing to do with principle. Lady Salisbury, the prime minister's wife, blamed that willfulness said to be characteristic of the Churchills throughout history. Whatever Lord Randolph's motives, he had clearly expected Lord Salisbury to refuse his resignation. Instead, the prime minister called his bluff. Lord Randolph's career was destroyed in an instant. A generation later, Winston was confident that this time round it was a Cecil who was about to self-destruct and a Churchill who would endure.

About an hour into his verbal jag, Churchill began to calm down. Had he simply worn himself out? The lull in the tempest allowed Macmillan to make the case that in spite of anything the prime minister said the resignations would do the government and the party much injury.

When Macmillan left, Churchill pretended to Colville that nothing of importance had been spoken of. He coyly claimed to have no idea why Clemmie had made such a fuss. Nevertheless, in important respects Macmillan's visit did seem to have an impact. Only now did Churchill begin to accept that Salisbury's resignation could have grave consequences.

Churchill moved at once to try to bind Eden and Butler to

him. On the night of July 16 he composed a telegram to Eden that mocked Salisbury as "this stickler for precise etiquette" and questioned the sincerity of his vaunted principles. He accused his opponent of being one of those men who "compound the sins they are inclined to by banning those they have no mind to." Salisbury's behavior the previous summer, when he had been a principal player in the medical cover-up, gave Churchill license to speak of him in this manner. The high-and-mighty Salisbury had been perfectly willing to get in the dirt with Churchill when it suited him, but he took a very different attitude now that strict adherence to the rules best served his interests. Churchill assured Eden that when Salisbury complained of constitutional improprieties he was arguing on very weak ground.

That Churchill was more worried than he let on is suggested by his behavior with Butler on July 19. Previously, Churchill had been adamant that he had felt no obligation to check in with the Cabinet before he approached Molotov. Now, he attempted to bully the chancellor into accepting an untruthful account that suggested that Churchill had really tried to consult the Cabinet after all. Pinocchio's nose grew longer as he claimed to have expected Butler to summon a Cabinet on receipt of the first telegram. It continued to grow when he said that he had assumed Butler's suggested edits were the product of Cabinet consultation.

In Churchill's intimidating presence, Butler made no protest. Still, he was unwilling to lie for Churchill again or, worse, to be cast as the fall guy. Two days later, Butler politely but vigorously disputed him in writing. Aware of what Churchill might yet be capable of, he took care to conclude with a gentle warning, pointing out that he was also sending a copy of his rebuttal to the Secretary of the Cabinet "so that it may be on record." If Churchill persisted, Butler would ask that the document be circulated to the full Cabinet.

Finally, Eden turned to the Americans for help with the old man. He asked the US representative in Geneva, Under Secretary of State Walter Bedell Smith, to come to London after the conference. Explaining that he was "gravely concerned" about Churchill's solitary pilgrimage to meet Malenkov, Eden enlisted Smith to try to talk him out of it. Churchill was fond of Smith, who had served as Eisenhower's chief of staff during the war. Nevertheless, on Thursday, July 22, when Smith made his pitch in advance of a dinner party in his honor at Number Ten, he was entirely unsuccessful. Churchill turned aside Smith's arguments that Malenkov was not actually filling Stalin's shoes and that at the moment it was quite possible that Molotov might be the more important figure. He sent Smith back to Washington with a message for Eisenhower: "I am an obstinate pig."

After dinner, Churchill drew Eden aside. He made it clear that he expected Eden's support the following morning when the critical Cabinet meeting was due to take place in Churchill's room in the House of Commons. Following their terse little talk, Eden was so flustered that he forgot about his wife and went home without her by mistake. Churchill assumed that yet again he had bent the heir apparent to his will. Friday morning, however, Churchill was about to leave for Westminster when a note from the foreign secretary took him by surprise. Eden had composed the message in the middle of the night in a fit of agitation. He wrote that on reflection he would not be able to do as he had been asked. As far as the meeting with Malenkov was concerned, Churchill was on his own.

Ministers gathered at eleven. Three other items came first on the agenda. When at last Churchill reached the potentially explosive fourth item, he calmly asked the Cabinet to approve his draft telegram to Molotov. The telegram, which he read aloud, suggested that before the British made a formal proposal he and Molotov should agree on a date and place for the meeting. Churchill named early September in Bern, Stockholm, or Vienna.

No sooner had he finished reading than Salisbury jumped in. His face "white and tense," Salisbury spoke from notes. He expressed the hope that "no message need be sent now leading up to a firm proposal for bilateral talks." But instead of focusing on the question of Cabinet consultation, as Churchill and the others had been expecting, Salisbury took a different tack. At the time of the previous

Cabinet it had looked as if he had no chance of persuading his colleagues to join him in trying to stop Churchill. Even ministers who agreed that they ought to have been consulted about the message to Molotov were unlikely to raise a fuss now that Churchill had Eisenhower's third telegram in hand. In the past ten days, Salisbury had had a chance to completely rethink his strategy. His comments of recent months about American recklessness had touched a chord with the Cabinet. Today, he began by recalling that initially it had been the constitutional implications of Churchill's actions that had chiefly concerned him.

"On reflection," Salisbury continued, "I find the international repercussions even more disturbing. Some people think Russia is now the greatest threat to peace. I don't. I think the main danger is from the United States." Not for the first time in this setting, Salisbury expressed concern that the Americans might be tempted to bring matters to a head while they still possessed overwhelming atomic superiority.

Salisbury's remarks tapped into strong Cabinet sentiment that the Americans had been playing a dangerous game in Indo-China and elsewhere. "During this period," he went on, "the supreme object of policy should be to preserve the unity of the West. How can we expect the US to respect that if we approach Russia without prior consultation?" If the British acted alone, Salisbury asked, was there not a risk that they would thereby encourage the Americans to pursue independent policies and to take less account of London's views?

In short, Salisbury was no longer attacking Churchill for having flouted the rules. He was calling him to task for having risked Britain's ability to restrain the United States. It was a potent argument and suddenly he seemed to have a real chance of defeating the prime minister on the central issue of Churchill's postwar premiership. Churchill could no longer sneer that Salisbury was concerned with mere questions of etiquette while he was engaged in the rather more important business of saving the world. Now, Salisbury also claimed to be trying to prevent a third world war. In closing, Salisbury urged the Cabinet not to allow Churchill to go forward with two-power talks. Churchill had boxed ministers in politically by acting first and informing them later, but Salisbury thought he saw a way out. This very morning Moscow had issued a new attack on US policies. Salisbury proposed that Molotov be informed that in light of these statements the entire situation needed to be reviewed.

Churchill was caught off guard. Suddenly, he had a real contest on his hands. At first, he tried to recover by arguing that of course there had been prior consultation with the Americans. He ticked off examples and he read aloud from Eisenhower's third telegram.

Salisbury was unimpressed. He sniffed, "No word was said before the approach to Molotov."

"Much was said informally, though with no time factor, in Washington," Churchill countered. "I mentioned a bi-lateral reconnaissance as well as an eventual three power meeting. They knew what was in my mind."

At this point, Churchill abruptly shifted the ground back to the issue of Cabinet consultation, where he preferred to fight. The debate over the international repercussions of his behavior was far more dangerous to him. Salisbury allowed himself to be lured into further discussion of the constitutional issue, and before long he had lost his edge. Lest he manage to regain it, Churchill called for a vote. Salisbury fought him off and at length Eden suggested that they postpone a final decision to give them all time to digest that morning's blast from Moscow. Churchill again demanded an immediate vote, but the ministers were desperate to stall. It was decided that they would reconvene on Monday, and the session ended with Churchill and Salisbury trading bitter threats to resign if the other prevailed.

Referring to the setting of their next encounter, Churchill declared, "If it is to be my last Cabinet, I should like it to be at Downing Street, not here [i.e. the House of Commons]." Then he looked over at his antagonist and huffily asked, "Will that suit Lord Salisbury?"

"Certainly," Salisbury said, "if I am to be there."

"What do you mean 'if I am to be there'?" Churchill fumed. "Will you resign before that?"

Salisbury flashed a wry smile. "I only said 'if.' "

As the curtain dropped, there was a sense that compromise between these two was impossible. If on Monday the Cabinet blocked Churchill, he was pledged to resign and to tell the public why. If they let the prime minister respond as he wished to Molotov, Salisbury would go.

That weekend, Churchill presided over a "stag-party" at Chequers attended by several Cabinet ministers. Churchill and his guests laughed, told stories, and recited poetry. But for all the gaiety, an "air of crisis" hung over the occasion. To their colleagues' horror, both Churchill and Salisbury seemed ready to go to the extreme if they did not get their way. Neither appeared to care if he brought the government and the Tory party down with him. Whether it was Churchill or Salisbury who resigned, Monday looked to be a day of doom.

On Saturday evening, the unexpected happened. The Soviets delivered identical notes to the British, American, French, and Chinese embassies in Moscow to propose a thirty-two-power foreign ministers' conference on European security issues. Clearly, the proposal aimed to derail French ratification of the EDC. The Soviet note was a game-changer and it profoundly unsettled Churchill. How could he possibly go ahead with efforts to arrange a top-level meeting as long as the new proposal was in play? In his great eagerness to return to the summit, had he misread the tenor and timing of Molotov's July 6 message?

The fireworks everyone had been dreading on Monday never happened. Instead, a strangely subdued Churchill informed the Cabinet that the Soviet note, which had been made public, constituted a "new event." Churchill opined that it was clearly calculated to block French ratification of the EDC. He said he was satisfied that he could not go forward with his proposal for bilateral talks while this suggestion of a much larger foreign ministers' meeting was being publicly canvassed. Accordingly, he had prepared a new reply to Molotov's July 6 message. He suggested saying that his own proposal would be held in abeyance while the Soviet note was under consideration.

Churchill wanted the Cabinet to understand that he was not conceding that he had done anything wrong. "I don't regret my approach. It was consistent with what I have said in public." But this burst of defiance quickly died out. What was the point? The battle was over, though in the end it was not Salisbury who had shot down Churchill's summit proposal, but the Soviets. Thus far in the meeting, Salisbury had remained silent. But when Churchill suggested that the Cabinet approve the substance of his message to Molotov, its precise wording to be settled later with Eden, Salisbury piped up that he would be happy to assent.

In the days that followed, Churchill wore his heavy heart on his sleeve. He seemed exceptionally old and confused; he alternated between sitting in silence and rambling on about nothing; he looked at times as if he might be about to suffer another stroke. When Parliament adjourned at the end of the week, the table talk in political London was that Churchill would have handed over power by the time they reassembled on October 19.

But then, on the very first day of the eleven-week recess, the situation shifted dramatically again—at least to Churchill's eye. He was at Chartwell on Saturday night when a new note from Molotov was brought in. Molotov expressed astonishment that Churchill seemed to think that the proposed thirty-two-power conference ruled out top-level talks. Again, Churchill went into overdrive, though the Foreign Office advised that Molotov was probably just trying to preclude accusations that the Kremlin had killed the British peace proposal. Churchill moved at once to resume the Cabinet debate where it had broken off. Salisbury had come at him with a powerful new argument. Though Churchill had managed to sidetrack him, Salisbury was sure to reprise his strongest points when the contest resumed. Cabinet members had already begun to leave town, so instead of calling a meeting at Downing Street, Churchill dictated a fresh set of arguments for bilateral talks and sent them around on August 3.

In the knowledge that Salisbury's argument about the American danger was weighing heavily with ministers, Churchill hijacked it. Churchill did not disagree that the Americans might be tempted to bring tensions to a head while they still had overwhelming atomic superiority over the Soviets. He, too, saw the American danger during the next two or three years as real and pressing. But where Salisbury cited that danger as a reason to avoid two-power talks, Churchill argued that it made those talks the more essential.

Salisbury took two weeks to respond. When he did (copying his remarks to Eden, Butler, Macmillan, and Crookshank) it was evident that his patience had worn thin and that he was determined to put an end to Churchill's summit dreams once and for all. This time, Salisbury said the unsayable. For nine years, Churchill had been fueled by the belief that if only he could return to the summit he could head off another world war. The particular men he wished to see and the tactics he planned to employ had changed over time. What had never altered was his confidence that summit talks were the best way to achieve a lasting peace and that his unique gifts, insights, and personal stature made him the best, no, the only figure in the West to conduct them. In 1954, as in 1945, Churchill had no doubt that face to face with the Soviet leadership he would be able to accomplish what no one else could. That clear conviction had helped him spring back from electoral defeat, two strokes, and other trials. Now, a very fed-up Salisbury bluntly informed Churchill that he had been operating on the basis of "an illusion."

Salisbury argued that summit talks were destined to fail and that not even someone as "pre-eminent" and "persuasive" as Churchill would be able to achieve anything by a personal meeting with Malenkov. Salisbury was not just attacking Churchill's core ideas. He was undermining his justification for clinging to power long past the stage when most people thought he ought to have gone. He was puncturing the myth of his indispensability. If the defining policy of Churchill's postwar premiership was indeed based on an illusion, what reason did he have to remain in the job? In closing, Salisbury renewed his threat to resign if the Cabinet sided with the prime minister on this.

Salisbury's attack seemed hardly to touch Churchill; at least, Churchill behaved that way. On August 21, he replied that he planned to go forward and warned Salisbury against being too rigid in his views. (Churchill was far from the first person to accuse a Cecil of lacking flexibility.) Churchill wrote: "In peace or war action is determined by events rather than by fixed ideas. One is fortunate when one has the power to decide in accordance with the factual circumstances of the day or even of the hour. I always reserve to myself as much of this advantage as I can get. Not only does one thing affect many others, but their proportions alter in an ever-changing scene." Why, in short, take a position before Salisbury saw where things stood in a few weeks' time?

Churchill expected to summon the Cabinet to vote on his proposal as soon as the Western allies formally rejected the idea of a thirty-two-power conference and the French ratified the EDC. He thought both matters would be out of the way by the end of the month or early September at the latest. Two days after he wrote to Salisbury, however, a crater opened before him. Following three years of endless delays, the French would not be ratifying the EDC after all. Western diplomats were going to have to develop a substitute plan to rearm West Germany and bring the Germans into Western Europe. Churchill gruffly acknowledged that until a new plan was agreed upon (a process that could take months, perhaps even years) he could not renew his proposal.

The day after Churchill heard about the French, he wrote to Eden that the September handover was off. He had no intention of abandoning his post during the present world crisis. As he prepared to lob this grenade into the Eden camp, he also had reason to expect quite an explosion at home. No doubt his wife would be as unhappy as his political heir about his determination to battle on. So he reprised the stunt that he had recently used with the Cabinet and with Eisenhower. He sent off the letter first and told Clementine about it later.

"It has gone," Winston reported on August 25 when he showed her a copy of what he had written to Eden. "The responsibility is mine. But I hope you will give me your love."

Churchill's eightieth birthday was fast approaching, and both houses of Parliament had commissioned the artist Graham Sutherland to paint his portrait. The picture was to be their joint gift to the prime minister, who agreed to sit for sketches on August 26. At Chartwell that afternoon, Sutherland lunched with the Churchills. Afterward, the men walked down to the painting studio together. There Churchill climbed onto a wooden platform which stood before a wall densely covered with his own framed paintings. Some of the pictures on display had been completed in Italy in 1945, after Churchill was hurled from power.

Sutherland, who had never met Churchill before this day, was encountering him at another moment of decision and defiance. Churchill had written to Salisbury and Eden. He had revealed his plans to Lady Churchill. Tomorrow, he intended to notify the Cabinet in person.

As he dropped into an armchair, he inquired whether Sutherland planned to portray him as Churchill the cherubic or the defiant. But it quickly became apparent that the sitter had his own firm ideas about how he wished to appear.

Churchill's lower lip jutted out. His eyes glowered. His head tilted challengingly.

XIX

THE "R" WORD

Westminster Hall, November 1954

Westminster Hall rocked with affectionate laughter when Churchill used the "R" word.

"This is to me the most memorable public occasion of my life," he told the 2,500 celebrants—members of all parties and both houses of Parliament, and their wives—who had gathered on November 30, 1954, to salute him on his birthday. "No one has received a similar mark of honor before. There has never been anything like it in British history and, indeed, I doubt whether any of the modern democracies abroad have shown such a degree of kindness and generosity to a party politician who has not yet retired . . ."

For more than a decade, people had been asking when Churchill planned to retire. Could he last to the end of the war or would he have to hand over to a younger, stronger man? By defeating the Nazis, had he not earned the right to a quiet, happy retirement? As the new Labour government was likely to be in for at least ten years, would it not be wise to accept that at his age he almost certainly could never be prime minister again?

Would he go after the Fulton speech? At the end of the first year to eighteen months of his new postwar government? At the coronation? As soon as Eden recovered from surgery in the US? When the Queen returned from her Commonwealth tour? In September 1954? By the time Parliament reassembled after the summer recess?

At his eightieth birthday?

On the eve of the tribute, Salisbury had predicted that rather than see the occasion as "just the moment for closing a great career," Churchill was "much more likely" to be confirmed in the opinion that he had "never been more necessary to the country."

Both main speakers on the day's program, Attlee and Salisbury, in their capacities as opposition leader in the Commons and leader of the Lords, paid homage to a giant whose place in history was assured. Salisbury said tenderly and movingly that it had been the privilege of the present generation to have seen and known Churchill for themselves. "That is something, I believe, for which we shall always be envied by those who come after."

Churchill lapped up the praise. Nevertheless, he made it clear that for all the talk of history he was not ready to be relegated to the past. "Ladies and gentlemen, I am now nearing the end of my journey," he said before he added, "I hope I still have some services to render."

Seated behind Churchill were ministers who would have been relieved to see him tender his resignation without delay. Churchill knew how they all felt. But, as ever, he preferred to seize on anything that might be construed as encouragement.

And what could seem more encouraging than the thunderous applause, the heartfelt testimonials, the flow of presents large and small from every corner of the nation? Churchill was particularly impressed by the nearly quarter of a million donations, ranging from less than a shilling to ten thousand pounds, to the Winston Churchill Eightieth Birthday Fund.

The only black moment was the public unveiling of the Sutherland portrait. Churchill, who had already seen the finished painting, felt that the artist had betrayed him by depicting too vividly the ravages that time and illness had wrought. Lest her husband spoil the day, Clementine Churchill counseled him beforehand to pretend that he was pleased. She later quietly burned the picture because he hated it so.

In the afterglow of the all-party tribute, Churchill wrote again to

Eisenhower of his abiding wish that they might yet meet Malenkov together: "It is in the hope of helping forward such a meeting that I am remaining in harness longer than I wished or planned." As far as Churchill was concerned, events in Europe since France's failure to ratify the EDC had confirmed the wisdom of his refusal to hand over in October. Thanks to Eden's deft diplomacy that autumn, agreements to resolve the German problem had been reached more quickly than anyone had expected. The Allied occupation was to end and West Germany to become a sovereign state and full member of NATO. Churchill was optimistic that the nine powers involved would be able to ratify the agreements early in the new year. After that, the way to the summit would be open.

Churchill reported to Eisenhower that when he had had his last audience with the Queen, she had spoken of the pleasure with which she would welcome a state visit by him to London. He went on, "This might be combined in any way convenient with a top level meeting."

Eisenhower moved quickly to dash Churchill's hopes. The president pointed out that ratification was likely to take a good deal longer than Churchill suggested. He noted that even if the accords went through, the Soviets, displeased by West Germany's entry into NATO, would probably "play tough" for a while at least. In any case, Eisenhower continued to believe that the top men ought not to meet until the foreign ministers had had a chance to lay the groundwork. Under the circumstances, he did not expect even second-level talks in the near future. "So, I am bound to say that, while I would like to be more optimistic, I cannot see that a top-level meeting is anything which I can inscribe on my schedule for any predictable date."

This was devastating, not because Eisenhower had again refused, but because of what he said about time. If the president had his facts right, and he seemed to, it would be many months before the Big Three could be reunited. Quite simply, Churchill did not have many months left in office. Through the years, Churchill had often warned of the shortness of time. Now, the old man sensed that time—his time—had just run out. Churchill was not alone in grasping the import of Eisenhower's December 14 telegram. When the message came in, Eden was about to go off to Paris for defense talks. Convinced that Churchill no longer had even the flimsiest excuse to cling to power, Eden asked to see him as soon as he returned. But when Eden arrived home on December 19, he took to his bed with a chill. As a consequence, he missed the next day's Cabinet when Churchill asked Woolton, Butler, and Stuart to stay behind afterward. To their surprise, Churchill proposed that they consider having a general election in the spring. As he did not plan to lead in a new election, the ministers understood him to have suggested that he might step down early in 1955.

It was one thing for Churchill to speak of these matters to Woolton and the others, and quite another to have to face Eden's "hungry eyes." The younger man no doubt would press for dates and specifics, and Churchill was far from ready to talk in those terms. Work and life were interchangeable to him. He was not anxious to sign what he bitterly described as his own death warrant. For the moment, Eden's illness suggested a convenient, if temporary, way out. Today was Monday. Due to leave town on the Thursday, Churchill told Eden on the phone that there would not be time to meet before Christmas. But Eden persisted and, though he had developed a fever, he went to Churchill the following evening. Clementine Churchill expressed alarm about Eden's sickly color. Churchill sneered that it looked as if he had been living too well in Paris.

Some men seek to bribe the grim reaper, others to trick him. Churchill tried intimidation. "What do you want to see me about?" he demanded of Eden in what the latter described in a diary entry as "his most aggressive tone." When Eden pushed for a definite date, Churchill pushed back. Far from speaking of an early election, he insisted he really need not go until the end of June or July. Eden protested that that would be too late. This prompted Churchill to play a favorite mind game. Macmillan, whom Churchill had recently appointed minister of defense as a reward for his successes at housing, and as part of a broader Cabinet reshuffle, had just accompanied Eden to Paris. Now, Churchill asked Eden how he got along with Harold.

"Very well," Eden replied, taking the bait. "Why?"

"Oh," said Churchill suggestively, "he is very ambitious."

In recent months, Churchill had tried as much as possible to face his attackers separately, the better to play them "one against the other." This evening, he grudgingly agreed to meet the next day with a group of Eden's choosing. Ostensibly, the purpose of the session would be to consider election dates.

When Eden, backed up by the usual suspects, Salisbury, Macmillan, Butler, Woolton, Stuart, and Crookshank, confronted Churchill on the afternoon of December 22, everyone knew that the real date in question was that of the old man's retirement.

It did not take long for Churchill to explode at Eden: "I know you are trying to get rid of me."

Pointedly, no one contradicted him.

Churchill said he refused to be hounded from office. He reminded the ministers that it was up to him to go to the Queen and hand her his resignation, but he vowed not to do it. He conceded that if they felt strongly they could always force his hand. Were a significant number of them to resign, an election would be inevitable.

Churchill added menacingly, "But if this happens, I shall not be in favor of it and I shall tell the country so."

Eden and his supporters left with their tails between their legs. If Churchill carried out his threat, he would badly, perhaps fatally, damage the party in both the House and the country. The leader Eden insisted was gaga had single-handedly outfoxed them again. Eden was "in despair." Stuart opined that the meeting had been painful but necessary; Churchill simply had had to be told that he could not persist in a course of "such utter selfishness." But had the prime minister taken their point? Stuart feared not; he had the uncomfortable feeling that Churchill meant to remain in office until he died or until Parliament ended in 1956.

Later in the day, Churchill merrily suggested another possibility. As Eden and some of the others seemed so intent on an early election, he proposed to give them what they wanted—with one important difference. Churchill phoned Woolton to say that the size of the Winston Churchill Eightieth Birthday Fund demonstrated his high electoral value and that perhaps he should be the one to lead in an early election. He asked the horrified party chairman to look into it.

As Parliament rose on December 23 and Churchill went off to Chequers, it was clear that he had had the best of his face-off with the mutineers. While Churchill was in the country and while Eden (temporarily and not a little gratefully) withdrew from the fray, Salisbury, Macmillan, Stuart, and Butler reconvened at the home of Harry Crookshank. Seven years after this particular brain trust had first assembled there to discuss this very problem, they still lacked a better idea of how to force Churchill out.

The difference in the winter of 1954–5 was that not even the most devoted Churchillians any longer questioned that there were legitimate concerns about the prime minister's ability to carry on. Norman Brook judged that, though Churchill could still rise to the great occasion by a sheer act of will and the use of amphetamines, he no longer had the energy to grapple with the day-to-day demands of the premiership. Jock Colville noted that with each passing month, Churchill's powers evaporated a bit more. Was Churchill still the man to face the Soviets? Absolutely not, the Edenites insisted. Colville admitted he just did not know. On icy winter nights, as Churchill and Colville played bezique and dined together, Churchill spoke of being "tired of it all" and having "lost interest" now that his hopes of a Big Three meeting had been deferred. A final decision seemed to be at hand; but could he bring himself to make it?

When Churchill refused to retire in 1945, his decision had flowed from everything that was essential to his character; so had his subsequent decisions to fight on. At the beginning of 1955, the decision that confronted Churchill was different, harder. This time, rather than ride the wave of his obstinacy, he had to overcome it. He had to crush his lifelong refusal to accept defeat. He had to conquer the primal survival instinct that had allowed him to spring back so many times before. This time, Churchill's battle was not really with Salisbury, Eden, Eisenhower, or any other antagonist. It was with himself.

At length, Churchill thought he might leave office at the Easter recess. First, he mentioned the date to Colville. Then, he broached it in conversation with Butler. Finally, on February 1, 1955, Eden came to Number Ten. The foreign secretary was preparing to go to a conference in Bangkok and he wanted to know Churchill's plans. After sending for a calendar at Eden's suggestion, Churchill mentioned Easter. But he asked Eden to keep that date a secret so that he could tell the Queen quietly and avoid a fuss. Eden was pleased by how smoothly the meeting had gone.

Whether Churchill could be trusted to follow through was another matter. On February 8, reports from Moscow suggested that the post-Stalin succession struggle was far from finished. Malenkov had fallen. Bulganin and Khrushchev had taken his place. Whose bones would fly out next was anyone's guess. Directly, Churchill asked to see Eden and Butler to coordinate the date for the Budget and the timing of the handover. On the evening of February 14, Eden emerged from Number Ten convinced that he finally had a solid commitment.

A two-day defense debate loomed in the House of Commons at the beginning of March. More than half a century after Churchill had first spoken in the House, he planned to use the occasion to deliver his last great address there. He lavished twenty hours on the composition of a forty-five-minute set-piece speech, a meditation on global statecraft in the age of the H-bomb. When he ordered up the necessary medical stimulants in advance of his performance, he told Moran that before he left office he intended to make it clear to the world that he was still fit to govern. He wanted people to understand that he was not retiring because he could no longer carry the burden but because he wished to give a younger man his chance.

Slapping the sides of the dispatch box as he addressed the House on March 1, 1955, Churchill observed that a quantity of plutonium less than would fill that very box would suffice to produce weapons capable of giving world domination to any great power which alone possessed it. "What ought we to do? Which way shall we turn to save our lives and the future of the world? It does not matter so much to old people; they are soon going anyway, but I find it poignant to look at youth in all its activities and ardor and, most of all, to watch little children playing their merry games, and wonder what would lie before them if God wearied of mankind."

His account of the policy of defense through deterrents was lucid and effective. His phrases sparkled. His voice throbbed as strongly at the close as it had at the start. At one fell swoop Churchill reasserted his gift of oratory and his mastery of the House. Even the *Daily Mirror*, which had been calling for the prime minister's resignation on grounds of senility, admitted that the performance had been Churchill "at his very best." Afterward, Soames assured his father-in-law, who was noticeably out of breath, that if he never made another speech in his life this had been "a very fine swan song."

Unexpectedly, the coda was still to come. Mission accomplished, Churchill did not plan to speak on the second day of the defense debate, only to sit beside Macmillan on the Tory front bench. But he was stung to interrupt when Aneurin Bevan insinuated that he had been dithering in the business of securing a top-level conference and that he had permitted himself to be at the beck and call of the Americans.

Churchill flung back: "It is absolutely wrong to suggest that the course which we have followed here has been at the dictation of the United States." The day before, Churchill had spoken with his notes laid out before him. Every element of his performance had been scripted. Nothing had been left to chance. By contrast, today's intervention was entirely off the cuff. Macmillan trembled for fear of what Churchill in his fury was about to say. Was he about to undo all of yesterday's fine work? Was this to be a rerun of the debacle of April 5, 1954? Equally nervous, Soames chewed his thumb until it bled.

Churchill spoke of his efforts to arrange a meeting with the Soviets after Stalin's death, of his original plan to meet Eisenhower in Bermuda, and of the real reason he had been unable to go. "I was struck down by a very sudden illness which paralyzed me completely, physically," he said as he ran his right hand down the right side of his body, from shoulder to knee. "That is why I had to put it all off."

He talked of Eisenhower's refusal to join him in making a summit proposal and of his own attempt to arrange a bilateral meeting. His references to the tumultuous events of the previous summer, after he sailed home from the US, caused stomachs to churn on the government bench. The Cabinet crisis had petered out without the potentially damaging details having become widely known. Was Churchill about to punish his party by revealing Eden's reluctance and Salisbury's intransigence? Would he acknowledge that the Cabinet had nearly broken up?

To his colleagues' vast relief, Churchill "slid successfully past these traps." He laid the blame for the failure of his Soviet initiative on the Kremlin's efforts to block ratification of the EDC. Now, he suggested, it remained only to wait until the NATO agreements were ratified. Once that process was complete, "any Government who are responsible at that time" would be free to meet the Soviets. Most of his audience was still reeling from the first public acknowledgment that he had had a stroke, but a careful listener might have caught a hint that by the time the road reopened Churchill himself would no longer be in power.

Churchill was exceedingly pleased with himself afterward. The two-day debate had allowed him to double-dip, both to prove that he had lost none of his old razzle-dazzle and to defend his reputation in the eyes of history. When he discussed the debate with his doctor, he even seemed capable of jesting about his departure from office. He assured Moran that he had no intention of going back on his date with Eden and that he was not considering a comeback.

Then he added slyly, "At least not yet."

Picturesque in a red velvet brocaded jacket and matching slippers, Churchill was full of jokes about retirement when he entertained the Edens at dinner on Tuesday, March 8. There was much ebullient table talk of the vacation he planned to take at Easter. Following the handover, he and Clementine were to go to Sicily with Bobbety and Betty Salisbury. They were all to stay at the deluxe San Domenico Hotel in Taormina, on the island's east coast. Over lunch on Wednesday, Eden expressed confidence to Macmillan that Churchill really meant it this time. On the Thursday, Clementine Churchill gave Clarissa Eden a tour of Number Ten in anticipation of the Edens' taking up residence there in less than a month.

On the Friday, Churchill tested the new Rolls-Royce he contemplated buying as a retirement gift to himself. En route to Chequers, he spontaneously stopped off at the London Zoo to visit the lion, Rota, and the leopard, Sheba, he kept there. If retired life was going to be like this, it could not be so bad; could it?

Later, he and Colville were playing bezique at Chequers when some official papers were brought in. Churchill read both the telegram from Sir Roger Makins which had been delivered earlier at the Foreign Office and Eden's covering minute. Neither document seemed to make much impression at the time. The men completed their card game and, at length, Lord Beaverbrook joined them for dinner. Only when Beaverbrook had gone did Churchill take a second, longer look at what the ambassador and the foreign secretary had written. When he finished reading he informed Colville that all bets were off. He had changed his mind about April. At the faintest flicker of encouragement, Churchill's instincts had kicked in.

That flicker had been very faint indeed—if there was a flicker at all. Makins's telegram sought London's response to the American plan for spurring France to approve the European security agreements. The White House, which at this point was unaware that Churchill had consented to leave office before Easter, suggested that Eisenhower might come to Paris on May 8, the tenth anniversary of V-E Day. His visit would provide an opportunity to ratify the agreements alongside Churchill, Adenauer, and President René Coty of France. Makins reported that Eisenhower had also suggested that while he was in Paris they could "lay plans for a meeting with the Soviets in a sustained effort to reduce tensions and the risk of war."

The phrase electrified Churchill. He read it as an indication that Eisenhower was finally ready to participate in top-level talks. He saw it as a last-minute reprieve. He transformed it into a lifeline to save him from having to retire. It referred neither to a Big Three meeting nor to Eisenhower's willingness to attend, but from then on, Churchill confidently spoke and acted as if it did. On the basis of it, he informed Colville that he had decided to remain in office and to meet the Soviets with Eisenhower.

When Colville pointed out that no suggestion of a heads of government meeting had been made, Churchill was brusquely dismissive. Nor would he listen to Clementine Churchill when she too sought to tamp down his enthusiasm. The Houdini of British politics went to bed that night convinced he still had some rope left after all.

In the morning, Churchill wrote to inform Eden that everything had changed. He exulted that this was "the first time" that Eisenhower had responded positively to his appeals. He decreed that the US proposal of leaders' talks which Eisenhower would attend "must be regarded as a new situation which will affect our personal plans and time-tables." He acted as if the angels had sung though only he seemed to have noticed. "The magnitude of the Washington advance towards a Top Level meeting is the dominant fact now before us."

Churchill was aware that Eden was inclined to call a May election to take advantage of Labour disunity, but he warned that that plan would have to be dropped. There must be no suggestion that considerations of party advantage had been permitted to override the quest for world peace. The people would not stand for it.

Eden was already in high hysteria when he received Churchill's message at lunchtime on Saturday. Makins's telegram had plunged him into "full crisis" the day before, when he realized that Churchill would no doubt use Eisenhower's Paris visit as an excuse to cling to office. After all these years, Eden knew his man. But even he had not foreseen that Churchill would go so far as to read a change of heart about summit meetings into Eisenhower's proposal. Outraged at what he took to be a nasty personal dig, Eden replied that he was not aware that anything he had done in his public life "would justify the suggestion that I was putting Party before country or self before either."

Eden's nerves worried Salisbury, who conferred with him several times in the course of the weekend. Along with Butler, Salisbury had agreed to join Eden in resigning if Churchill tried to wiggle out of the April date. For the moment, however, Salisbury was confident that Churchill could be stopped. The present crisis was entirely different from the one during the summer. Then Churchill had actually managed to wheedle an okay out of Eisenhower, and there had been a real possibility that he might have persuaded the rest of the Cabinet to back him on bilateral talks. This time, one needed only to examine the Washington telegram to see that Eisenhower had not proposed to meet the Soviets. Conveniently, Churchill in his exuberance had already circulated copies to the full Cabinet. As far as Salisbury was concerned, the most important thing their side could do on Monday was to coolly point out that the emperor had no clothes. But was Eden up to it? Salisbury confessed to Macmillan that he feared Eden might not be able to stand the strain.

At the Monday Cabinet, Churchill and the ministers talked past one another for the better part of an hour. The ministers concentrated on the pros and cons of an Eisenhower visit in terms of the ratification issue and of Tory election plans. Bewilderingly to some of his colleagues, Churchill kept coming back to the prospect of a summit meeting. The way he told the tale, there was a good deal more to Eisenhower's trip than an effort to prod the French. Churchill had already been to Buckingham Palace for permission to invite Eisenhower to London, so talks with the Soviets might begin at once.

Finally, Salisbury doused cold water on the prime minister's excitement. "I don't think Eisenhower contemplates a top level meeting. He never has. As a constitutional monarch, he will confine himself to formal acts. Almost certainly, he thinks in terms of a foreign ministers' level meeting." Salisbury went on to argue that in any event the Conservatives must retain the option of calling a May election, in which case a presidential visit then would not do. The ministers discussed asking Eisenhower to put off his trip until June. When Churchill spoke, he made it sound suspiciously as if he planned still to be in office then.

"Does that mean, Prime Minister," Eden asked slowly and unemotionally, "that the arrangements you have made with me are at an end?"

The question took Churchill aback. "This is a new situation," he mumbled. "I should have to consider my public duty." Always at such moments he spoke in terms of duty.

Eden's brittle veneer of calm began to crumble. "Does that mean, Prime Minister," he challenged, "that if such meetings were to be held there is no one capable of conducting them?"

Churchill replied, "It has always been my ambition—this is too great a national and international opportunity to yield to personal considerations."

Eden protested, "I have been Foreign Minister for ten years. Am I not to be trusted?"

Churchill took this as his cue to wax indignant and to cite his even lengthier public service. "All this is very unusual," he blustered. "These matters are not in my long experience discussed in Cabinets." He put Eden and the others on notice that the decision of if and when to leave office was his alone to make.

Salisbury tried to bring down the temperature, but his remarks had the opposite effect. He pointed out that not everyone present was aware of the arrangements Churchill and Eden had been speaking of. Only Churchill, Eden, Salisbury, Butler, and Macmillan had definite knowledge that Churchill had agreed to go in April. Woolton and Swinton knew some of it. Salisbury's suggestion that the full Cabinet be apprised of his retirement plans infuriated Churchill, who had just made it clear that this was a subject he did not wish to discuss.

"I cannot assent to such a discussion," Churchill thundered. "I know my duty and will perform it. If any member of the Cabinet dissents his way is open."

The Cabinet Room seethed with embarrassment. Churchill thought everyone seemed so pained because Eden had dared to raise the personal issue, but that was only part of it. The Washington telegram appeared to have pushed Churchill over the line that separates optimism from self-deception. Perhaps he really did believe that Eisenhower had made a "new offer," but no one else did.

As the meeting broke up, Churchill asked to see Salisbury privately. When they were alone, Churchill said he hoped Salisbury still planned to go to Sicily with him. Salisbury said he did. Testing the waters further, Churchill asked, "It isn't against your principles to fly on holiday with a Prime Minister?" This was his arch way of stating that he did not intend to yield. Having already made one false step, Salisbury refused to be provoked. After all, at this point Salisbury really had no need to do anything but wait. Sooner or later, Churchill was bound to crash headlong into the mountainous fact that the "new offer" was a chimera.

The crash came that very day. Aldrich delivered a message from Washington that stated flatly that Eisenhower was not willing himself to take part in a meeting with the Soviets. That certainly would have seemed to be the end of that—at least, the Edenites thought so. But it was not quite enough for Churchill. Still scrambling to find a way out of an April handover, he insisted on going back to the Americans one more time. Again, he struggled mightily to wring a more hopeful meaning out of the words in front of him. He knew his interpretation was a stretch (he admitted as much to Clementine), but given the alternative he felt he had to try. He asked whether this latest Washington message referred exclusively to Eisenhower's upcoming visit. Did it necessarily rule out talks at a somewhat later date?

Churchill had his answer soon enough. Makins telegraphed that

Eisenhower was not contemplating an early top-level meeting. The president had merely thought that while he was in Europe, preliminary arrangements might be made for foreign ministers' talks to take place in October. On March 16, Churchill read the telegram aloud to the Cabinet. When the old man spoke the word "October," he made "a gesture of disappointment." Churchill seemed to accept that the struggle was at an end. He would retire in April after all.

The mood at Chequers that weekend was very somber. "It's the first death," Clementine Churchill reflected, "and for him, a death in life." Though she was relieved that the matter of Winston's retirement was settled at last, she shared his anguish nonetheless. He had confronted the abyss in 1915 and again in 1945. The difference in his eighty-first year was that there could be no "next time." Once he tendered his resignation, his political life would be over.

For ten years, the war lord had longed to "round off" his story by making the peace. In all that time, he had never lost faith in his Soviet strategy or in his unique ability to execute it. Now that the last prize had eluded his grasp, it galled him to leave the leadership of the Western alliance in the hands of men whose expertise and understanding he did not trust. Churchill raged that Eden had done more to thwart him and to prevent him from pursuing the policy he believed in than anyone else. Nor, in private, did he have kind words for Eisenhower. Churchill said he would have stayed on had there been a chance of a summit meeting. "But," he continued, "Ike won't have it. He's afraid—and there it is."

Sadly, the universal applause that had delighted Churchill at his eightieth birthday tribute had faded from his ears. The sense of peace he had begun to acquire in the aftermath of the defense debate also was gone. He bristled at the political gossip that Eden had finally managed to kick him out. He brooded that in an effort to ensure his departure his own party must be responsible for leaks to the press that Eden was to succeed him before Easter. He thrashed about miserably as the countdown to April 5 began: his last weekend at Chequers, the Edens' dinner party in his honor,

Queen Elizabeth's highly unusual visit to Number Ten on the eve of his resignation. Butler compared Churchill to a fish that had been hooked but had yet to be gaffed. Colville saw him as having almost been landed but still fighting to break free from the net.

On Sunday, March 27, Churchill was spending his final days at Chequers when he saw newspaper reports that Bulganin had commented favorably on the idea of a conference of the great powers. The following day, Eden said in the House of Commons that it would be best if such contacts began at the foreign ministers' level. Churchill had been complaining of precisely that approach since 1945. That night, he told Colville that he could not possibly hand over at such a moment just to gratify Eden's personal hunger for power. When Churchill threatened to call a party meeting and ask Conservatives to choose, Colville warned that that would make an unhappy last chapter to his biography. Did Churchill really want it to be written that at the end of his career he had brought down the party of which he was the leader? Colville assumed that Churchill's threats were nothing more than a late-night fantasy, which reflected the anguish with which he contemplated the approach of retirement. No doubt the prime minister would think better of it in the morning. Instead, the next day, Churchill sent Butler to notify Eden that the grim international situation made it impossible for him to retire after all.

When Churchill answered questions in the Commons that afternoon, it was evident that summit dreams were again dancing in his head. He said he hoped a leaders' meeting might yet be arranged. When a Labour member asked whether he thought he could get one in time to participate himself, Churchill answered coyly that the future was "veiled in obscurity." Afterward, at his weekly audience with the Queen, he disclosed that he contemplated postponing his resignation. When Churchill asked whether she would object, the young sovereign said no. Some people at the time believed that Elizabeth's father would have been less understanding and that matters would never have proceeded to this point had he still been alive.

That night of all nights, Churchill was due at the Edens' dinner party. He rang to say he had been delayed at Buckingham Palace and would be late. Clementine Churchill phoned separately to report that her neuritis had flared up and that she could not attend. Eden, who had been frantic since morning, interpreted Churchill's call as a very bad sign. Colville sent word to remind Eden that Churchill throve on confrontation. To make a scene of any kind would only fuel his determination to battle on. When Churchill arrived that evening, Eden bit his tongue and pretended that nothing was wrong. Strange to say, so did Churchill. He sweetly inquired how his niece had enjoyed her tour of Number Ten and spoke of being in Venice in June. What did the mixed signals mean? To the last minute, Churchill seemed not to know himself what he planned.

At half past six the next evening, Churchill summoned Eden and Butler to the Cabinet Room. The men had been rivals since the end of the war, but at this late date almost no one believed that Butler had ever had a real chance of supplanting the designated heir. Yet, when they joined Churchill on March 30, the prime minister invited Butler to sit on his right. The gesture seemed to suggest that there had been a last-minute upset and that after all these years Eden had been dealt a terrible blow. But Churchill quickly corrected himself. He signaled Eden to take the place on his right, and he seated Butler on his left. Had the incident been a slip? Or a final stab of cruelty at his political heir's expense?

For a moment, the three of them sat without speaking. Finally, Churchill broke the silence.

"I am going and Anthony will succeed me," he declared. "We can discuss details later."

In the morning, Churchill sent a message to the palace. April 5 it would be.

During the next four days, he alternated between groaning that he was not ready to retire and acknowledging that he could not keep others waiting, between sighing that he did not wish to live very long after he relinquished power and insisting that he would still wield great influence. Churchill had written in his biography of the 1st Duke of Marlborough (1936) that it is best for a warrior "to die in battle on the field, in command, with great causes in dispute and strong action surging round." But, he went on, that is not always possible. It had not been the lot of Churchill's ancestor to make such a perfect exit. Nor, he finally accepted that spring of 1955, was it to be his. Still, he did not want anyone to think that he had been dictated to or defeated by Eden, by the party, or even by age and failing health. That would be intolerable. To the end, he insisted that the timing and circumstances of his departure were his alone to choose. Even at this point, there was no individual or group in a position to force him to go. He had to surrender the seals of office of his own volition, and until the hour itself no one could be certain he really would.

Yet the momentum was steadily, relentlessly building. Shortly before midnight on April 4, Churchill emerged from Number Ten at the close of his farewell banquet, which had been attended by fifty guests including Queen Elizabeth and Prince Philip. Churchill wore black knee breeches, long black silk stockings, and his blue Garter sash. The Queen followed a few steps behind, the diamonds in her tiara flashing in the photographers' floodlights. Cameras clicked as Churchill bowed low before her and took the white-gloved hand she extended in farewell. As there would be no photographs inside the palace the next day, this image promised to be the one that would stand in history.

After the Queen's dark red Rolls-Royce drove off, Churchill went upstairs to his room. Still wearing his breeches and Garter sash, he sat in meditative silence on the edge of the bed for several minutes. That night, glasses had been raised. Pictures had been taken. And though (alarmingly, to some ears) Churchill had not spoken the "R" word in his after-dinner speech, the significance of the evening had been well understood by all. It would certainly seem as if there was no longer a way out of what faced him the next day.

Or was there?

As if he were still searching for an excuse to stay on, still struggling to reassure himself he was indispensable, he looked up and said, "I don't believe Anthony can do it."

The next day at noon, Churchill presided over the Cabinet for the last time. The showdowns, the maneuvers, the machinations ended abruptly when he announced, "I have decided to resign." After the full Cabinet posed for a group photograph, each minister came up to shake Churchill's hand and say a few words. Eden, Salisbury, Butler, Macmillan, and the rest were full of praise and affection for the leader they had long hoped to unseat. At a quarter past four, Churchill donned the frock coat and top hat he customarily wore at audiences with the Queen and went to the palace to surrender the seals of office. When he returned to Number Ten, a crowd was waiting.

"Good old Winnie!" they shouted.

His eyes filled with tears as he gave the V-sign. Then he disappeared inside and the door closed after him.

ACKNOWLEDGMENTS

My first and greatest debt is to Deborah, Dowager Duchess of Devonshire, who did so much to make it possible for me to tell this story. She laid out the terrain, provided key insights into many of the major players in the drama, and was generous in ways too numerous to list. She helped me begin to grasp just how complicated—and exciting—were the motives and relationships, the history and politics, of Churchill's world in these years.

I would like to thank the following people for their generous help and for their patience with my endless questions:

Lady Elizabeth Cavendish and Lady Anne Tree, for filling me in on the cast of characters and their tangled web of relationships.

Lady Soames, for speaking to me about her parents, Winston and Clementine Churchill, and especially for pinpointing the moment in 1945 when her father made the decision to fight on. Her discussion of the fraught issue of retirement; of her father's focus on the Russians during his final premiership; and of his 1953 stroke were all essential.

Sir Nicholas Henderson, for providing an eyewitness account of Churchill at Potsdam, for explaining what that conference meant to Churchill and for illuminating the characters of Eden and Butler, for both of whom he served as private secretary.

Lord Carrington, for telling me what it was like to be a member of Churchill's postwar government and for speaking of Churchill's situation after the war and of his relationships with Eden, Salisbury, and Macmillan.

Both Sir Nicholas and Lord Carrington talked to me about the significance of Eisenhower's mistake over Berlin, and Truman's refusal to come to London to consult with Churchill before Potsdam. It was my talks with them that made me realize that Potsdam was the place to start this story.

Lord Salisbury, for talking to me about his grandfather Bobbety Salisbury, and for clarifying aspects of his relationships with Churchill, Eden, and Macmillan, as well as of the Churchill-Cecil family relationship.

Hugh Cecil, for his insights into his uncle Bobbety Salisbury's character and for discussing him in the context of the Cecil family.

Lord Stockton, for speaking to me of his grandfather Harold Macmillan's role and of his relations with Churchill, Salisbury, and Eden.

Lord Norwich, for sharing his vivid memories of Churchill's visits with his parents, Lady Diana and Duff Cooper, and especially for his account of Churchill in Venice in 1951.

Also, two people who were no longer alive when I began this book, but whose influence was felt throughout:

The late Lady Lloyd, for countless hours of conversation in which she made the world of this book come alive for me. I first heard many of the names in this story listening to her.

And the late and wonderful Andrew, 11th Duke of Devonshire, who—almost ten years ago—put a copy of Volume I of Churchill's *The World Crisis* in my hands and said, "Start here."

Finally, I wish to express my gratitude to Helen Marchant, Lindsay Warwick, and Alex Harttung. And to the essential people who understood from the first what I wanted to do with this book and helped me every step of the way: my publishers and editors Jonathan Burnham, Gail Winston, Arabella Pike, Sophie Goulden, and Allegra Huston; and my agents Michael Carlisle and Bill Hamilton.

To my husband, David, my thanks for simply everything.

SOURCE NOTES

I. You Will, but I Shall Not

1 background on Churchill in Berlin and Potsdam: Lady Soames, Sir Nicholas Henderson, Lord Carrington, author interviews.

1 British party had swelled: Sir Alexander Cadogan, *The Diaries of Sir Alexander Cadogan* (New York: Putnam's, 1972), July 17, 1945.

1 "We have . . .": WSC, June 22, 1941, quoted in *New York Times*, June 23, 1941.

2 "bloodthirsty guttersnipe": Ibid.

2 "and every foot . . .": Winston S. Churchill, *Great Contemporaries* (London: Thornton, Butterworth, 1937), 63.

2 questioned the Russian soldier: *The Times*, July 17, 1945.

2 followed the Russian soldier: Lord Moran, *Diaries of Lord Moran* (Boston: Houghton Mifflin, 1966), July 16, 1945.

3 "I never think . . .": Quoted in Diana Cooper, *Autobiography* (New York: Carroll & Graf, 1985), 668.

3 "last pull up . . .": Violet Bonham Carter, *Champion Redoubtable: The Diaries and Letters of Violet Bonham Carter, 1914–1944*, ed. Mark Pottle (London: Orion, 1998), August 1, 1945.

4 turning away in disgust: *The Times*, July 17, 1945.

4 tested it first: Ibid.

5 "showdown": Prime Minister Churchill to Anthony Eden, May 4, 1945, *Foreign Relations of the United States: diplomatic papers: the Conference of Berlin (the Potsdam Conference), 1945* (Washington: United States Printing Office, 1945), vol. 1.

5 troubled by doubts: John Colville, *The Fringes of Power: Downing Street Diaries* (London: Weidenfeld & Nicolson, 2004), February 23, 1945.

5 Eisenhower's mistake on Berlin: Sir Nicholas Henderson, Lord Carrington, author interviews.

5 presses Eisenhower to take Berlin: Prime Minister to General Eisenhower, March 31, 1945, quoted in Winston Churchill, *Triumph and Tragedy* (Boston: Houghton Mifflin, 1985), 405.

5 "lost its former . . .": Quoted in Churchill, *Triumph and Tragedy*, 402.

6 sat in silence: Moran, *Diaries*, July 16, 1945.

6 saluting soldiers: Ibid.

6 "high as kites": Sir Nicholas Henderson, author interview.

6 "Force and facts": WSC to Anthony Eden, April 1, 1944, quoted in Martin Gilbert, *Winston S. Churchill*, vol. 7, *Road to Victory, 1941–1945* (Boston: Houghton Mifflin, 1986), 725.

7 "how much we have": Prime Minister Churchill to President Truman, May 6, 1945, *FRUS: Conference of Berlin*, vol. 1.

7 Truman preferred to wait: President Truman to Prime Minister Churchill, May 9, 1945, ibid.

8 confer first in London: Prime Minister Churchill to President Truman, May 11, 1945, ibid.

8 "ganging up": President Truman to Prime Minister Churchill, May 12, 1945, ibid.

8 "iron curtain": Prime Minister Churchill to President Truman, May 12, 1945, ibid.

8 surely it was vital: Ibid.

8 did not wish to put in writing: President Truman to Prime Minister Churchill, May 22, 1945, ibid.

8 wanted to see Stalin first: Joseph Davies to President Truman, June 12, 1945, ibid.

8 Churchill waxed indignant: Ibid.

8 Davies blamed Churchill: Ibid.

8 "placed not only . . .": Ibid.

10 Stimson luncheon: Henry Stimson Diaries, July 16, 1945, Yale University Library.

10 Stimson's insistence: Ibid.

11 "We will feel . . .": First Plenary Meeting, July 17, 1945, *FRUS: Conference of Berlin*, vol. 2.

11 nervous about facing: Charles Bohlen, *Witness to History* (New York: Norton, 1973), 226.

11 "I don't just . . .": First Plenary Meeting, July 17, 1945, *FRUS: Conference of Berlin*, vol. 2.

12 "of all hues": Cadogan, *Diaries*, July 18, 1945.

12 grown so accustomed: Record of Private Talk between the Prime Minister and Generalissimo Stalin after the Plenary Session on July 17, 1945, at Potsdam, PREM 3/430/7, The National Archives, Public Record Office.

13 The British ambassador: Archibald Clark Kerr Diaries, August 16, 1942 (Baron Inverchapel Papers), FO 800/300, PRO.

13 build a relationship with Stalin: Lady Soames, author interview.

13 "physically rather oppressed": Churchill, *Triumph and Tragedy*, 548.

14 conversation at dinner: Record of a Private Talk between the Prime Minister and Generalissimo Stalin at Dinner on July 18, 1945, at Potsdam, PREM 3/430/6.

14 background on Churchill-Eden relationship: Sir Nicholas Henderson, the

Dowager Duchess of Devonshire, Lord Salisbury, Lord Carrington, Lord Stockton, Lady Anne Tree, Lady Elizabeth Cavendish, author interviews.

14 "resolution, experience . . .": Quoted in Robert Rhodes James, *Anthony Eden: A Biography* (New York: McGraw-Hill, 1987), 265.

14 "Let Vyacheslav . . .": Quoted in *Molotov Remembers: Conversations with Felix Chuev*, ed. Albert Resis (Chicago, Ivan R. Dee, 1993), 190.

14 "lard-white": Sir M. Peterson to Mr. Bevin, March 22, 1949, in *British Documents on Foreign Affairs: Reports and Papers from the Foreign Office Confidential Print,* part 4, series A, *The Soviet Union and Finland,* vol. 6, *1949* (Bethesda, MD: University Publications of America, 2002), 266.

14 "a man of . . .": George Kennan interview, National Security Archive, George Washington University.

15 "no questions asked": Sir M. Peterson to Mr. Bevin, March 22, 1949, in *BDFA,* part 4, series A, vol. 6, 275.

15 Molotov's sleeping habits: James Stuart, *Within the Fringe* (London: Bodley Head, 1967), 130.

15 Stalin spoke of retiring on a pension: *Molotov Remembers*, 190.

16 interminable: Moran, *Diaries*, July 21, 1945.

16 study the report in full: Stimson Diaries, July 22, 1945.

16 rushed over to see Truman: Truman-Churchill Meeting, Sunday, July 22, 1945, 12:15 p.m., *FRUS: Conference of Berlin*, vol. 2.

16 spoke excitedly of the bomb: Moran, *Diaries*, July 23, 1945.

17 laid out the new situation: Field Marshal Lord Alanbrooke, *War Diaries, 1939–1945* (London: Weidenfeld & Nicolson, 2001), July 23, 1945.

17 Truman would never provide: Lord Carrington, author interview.

17 Stalin would shrug off: Sir Nicholas Henderson, author interview.

17 might yet prove to have been exaggerated: Alanbrooke, *War Diaries*, July 23, 1945.

18 sharpest exchanges: Eighth Plenary Meeting, Tuesday, July 24, 1945, 5 p.m., *FRUS: Conference of Berlin*, vol. 2.

18 "Fairy tales!": Ibid.

19 Churchill's dream: Moran, *Diaries*, July 26, 1945.

19 hoped to be back: Ninth Plenary Meeting, Wednesday, July 25, 1945, 11 a.m., *FRUS: Conference of Berlin*, vol. 2.

19 mood on the flight home: Lady Soames, author interview.

20 complained of heat and want of air: Nicholas Henderson, *Inside the Private Office* (Chicago: Academy Chicago, 1987), 20.

20 "a blessing in disguise": Quoted in Churchill, *Triumph and Tragedy*, 583.

20 stepping aside on Monday: Sir Alan Lascelles, *King's Counsellor: The Diaries of Sir Alan Lascelles*, ed. Duff Hart-Davis (London: Weidenfeld & Nicolson, 2006), July 26, 1945.

21 in war one: Alfred G. Gardiner, *Portraits and Portents* (New York: Harper & Row, 1926), 57.

21 "Unsquashable resilience": Violet Bonham Carter, *Winston Churchill: An Intimate Portrait* (New York: Harcourt Brace, 1965), 3.

21 as many arrows: Gardiner, *Portraits and Portents*, 57.
21 at least a decade: Henderson, *Inside the Private Office*, 21.
21 in for a generation: Harold Macmillan, *Tides of Fortune* (London: Macmillan, 1969), 286.
22 "You will, but I": Anthony Eden, diary entry, July 27, 1945, quoted in Anthony Eden, *The Reckoning* (Boston: Houghton Mifflin, 1965), 639.
22 food and a philosophic temperament: WSC to Lady Randolph Churchill, October 2, 1897, quoted in Randolph Churchill, *Winston S. Churchill*, vol. 1, *Companion, Part 2, 1896–1900* (London: Heinemann, 1967), 797.
22 Attlee arrives a day late: Telegram: Foreign Office to Terminal, July 27, 1945, FO 371/50867.
22 talks resumed: Record of Meeting between Prime Minister and Foreign Secretary and Generalissimo Stalin at Potsdam, July 28, 1945, at 10 p.m., PREM 3/430/9.

II. Face Facts and Retire

23 slept with the key: Sarah Churchill, *A Thread in the Tapestry* (London: André Deutsch, 1967), 86.
23 sleeping near a balcony: Moran, *Diaries*, August 2, 1945.
24 old music hall song: Churchill, *A Thread in the Tapestry*, 88.
24 "new boys": Cuthbert Headlam, *Parliament and Politics in the Age of Churchill and Attlee: The Headlam Diaries, 1935–1951*, ed. Stuart Ball (London: Cambridge University Press, 1999), August 1, 1945.
24 raucous cheers: Chips Channon, *Chips: The Diaries of Sir Henry Channon*, ed. Robert Rhodes James (London: Orion, 1996), August 1, 1945.
25 "the chorus of birds . . .": Arthur Booth, *British Hustings 1924–1950* (London: Muller, 1956), 228.
25 Churchill made intention clear at Chequers: WSC to Hugh "Linky" Cecil, July 29, 1945, quoted in Martin Gilbert, *Winston S. Churchill*, vol. 8, *Never Despair, 1945–1965* (Boston: Houghton Mifflin, 1988), 112.
25 his party's unanimous support: Moran, *Diaries*, August 2, 1945.
25 trio of Conservative heavyweights met: Anthony Eden, diary entry, August 1, 1945, and Halifax, diary entry, August 1, 1945, quoted in Gilbert, *Churchill*, vol. 8, 117.
25 background on Cranborne: Lord Salisbury, the Dowager Duchess of Devonshire, Lady Elizabeth Cavendish, Lady Anne Tree, Hugh Cecil, Lord Carrington, Sir Nicholas Henderson, Lord Stockton, author interviews.
25 "Your family has . . .": Lord Salisbury, author interview.
26 immensely attractive: the Dowager Duchess of Devonshire, author interview.
26 operate behind the scenes: Lady Anne Tree, author interview.
26 "objectivity": Sir Nicholas Henderson, author interview.
26 warmly acknowledged: Lord Salisbury, author interview.
27 "face facts and retire . . .": Cranborne to Paul Emrys Evans, January 19, 1946, Paul Emrys Evans Papers, British Library.

28 "knew he could bully . . .": Quoted in Andrew Roberts, *The Holy Fox* (London: Weidenfeld & Nicolson, 1991), 274.

29 Potsdam fails to produce settlement Churchill chased: WSC, August 16, 1945, in *New York Times*, August 17, 1945.

29 viewed Attlee warily: Sir Nicholas Henderson, author interview.

29 palpably shaken: Ibid.

29 seemed to have lost interest: Sir Nicholas Henderson and Robert Murphy quoted in Alan Thompson, *The Day Before Yesterday* (London: Sidgwick & Jackson, 1971), 30.

29 Halifax's visit to Claridge's: Lord Halifax, diary entry, August 1, 1945, quoted in Gilbert, *Churchill*, vol. 8, 117.

30 if only Eden had not been ill: Anthony Eden, diary entry, August 1, 1945, quoted in Rhodes James, *Anthony Eden*, 311.

30 "Some men need drink . . .": Quoted in Cynthia Gladwyn, *The Diaries of Cynthia Gladwyn*, ed. Miles Jebb (London: Constable, 1995), January 9, 1957.

30 swallowed a sleeping pill: Moran, *Diaries*, August 2, 1945.

30 Since Saturday: Ibid.

32 "gleefully anticipating": Lascelles, *King's Counsellor*, August 15, 1945.

32 the danger of a new war: WSC, August 16, 1945, in *New York Times*, August 17, 1945.

33 "a real masterpiece": Hugh Gaitskell, *The Diary, 1945–1956*, ed. Philip M. Williams (London: J. Cape, 1983), 18–19.

33 "brilliant moving gallant . . .": Clementine Churchill to Mary Churchill, August 18, 1945, quoted in Mary Soames, *Clementine Churchill* (Boston: Houghton Mifflin, 1973), 429.

34 background on Churchill marriage: Lady Soames, the Dowager Duchess of Devonshire, Lady Elizabeth Cavendish, Lord Norwich, author interviews.

34 soothed his bitterness: Hugh Cecil, author interview.

34 put him above their children's needs: the Dowager Duchess of Devonshire, author interview.

34 she would not hesitate: Clementine Churchill to WSC, April 6, 1916, in Mary Soames, ed., *Winston and Clementine* (Boston: Houghton Mifflin, 2001).

35 would grow annoyed: Lady Soames, author interview.

35 painful to be impotent and inactive: Martin Gilbert, *Winston S. Churchill*, vol. 3, *Challenge of War, 1914–1916* (Boston: Houghton Mifflin, 1971), 501.

35 less forgiving: Clementine Churchill to Mary Churchill, August 18, 1945, quoted in Soames, *Clementine Churchill*, 429.

35 Clementine regretted: Ibid.

III. Sans Soucis et Sans Regrets

37 flight to Milan: Moran, *Diaries*, September 2, 1945.

37 "an exhilarating drink": WSC to Lady Randolph Churchill, February 12,

1897, quoted in Randolph Churchill, *Winston S. Churchill*, vol. 1, *Youth, 1874–1900* (Boston: Houghton Mifflin, 1966), 315.

39 dinner at Villa La Rosa: WSC to Clementine Churchill, September 5, 1945, in Gilbert, *Churchill*, vol. 8, 137; Sarah Churchill, *Keep on Dancing* (New York: Coward, 1981), 137; Moran, *Diaries*, 315.

39 feared time would pass slowly: Churchill, *A Thread in the Tapestry*, 90.

39 "die of grief": Clementine Churchill quoted in Gilbert, *Churchill*, vol. 3, 473.

39 "knew everything and . . .": Winston S. Churchill, *Painting as Pastime* (New York: Cornerstone Library, 1950), 16.

40 first day of painting in Italy: Sarah Churchill to Clementine Churchill, September 3, 1945, quoted in Churchill, *Keep on Dancing*, 136; WSC to Clementine Churchill, September 3, 1945, in Soames, ed., *Winston and Clementine*; Moran, *Diaries*, September 3, 1945; Churchill, *A Thread in the Tapestry*, 91–3.

40 "voluptuous kick": Bonham Carter, *Winston Churchill*, 381.

41 "layer after layer": Churchill, *Painting as Pastime*, 18.

41 "I've had a . . .": Sarah Churchill to Clementine Churchill, September 3, 1945, quoted in Churchill, *Keep on Dancing*, 139.

41 He rejoiced: WSC to Clementine Churchill, September 5, 1945, in Soames, ed., *Winston and Clementine*.

42 no newspapers: WSC to Clementine Churchill, September 5, 1945, and September 13, 1945, in Soames, ed., *Winston and Clementine*.

42 glad to have been relieved of responsibility: WSC to Clementine Churchill, September 5, 1945, in Soames, ed., *Winston and Clementine*.

42 "Every day . . .": Sarah Churchill to Clementine Churchill, September 8, 1945, quoted in Churchill, *A Thread in the Tapestry*, 95.

42 "incongruous": WSC to Clementine Churchill, September 24, 1945, quoted in Gilbert, *Churchill*, vol. 8, 152.

43 bathes at Villa Pirelli: Gilbert, *Churchill*, vol. 8, 151.

44 he spoke of how certain: WSC to Clementine Churchill, September 24, 1945, in Soames, ed., *Winston and Clementine*.

44 his family understood: Lady Soames, author interview.

IV. Old Man in a Hurry

45 "a new man": Lascelles, *King's Counsellor*, October 26, 1945.

45 with a cold: WSC to Clement Attlee, October 6, 1945, quoted in Gilbert, *Churchill*, vol. 8, 156; *New York Times*, October 15, 1945; WSC to Sir James Hawkey, October 18, 1945, quoted in Gilbert, *Churchill*, vol. 8, 159.

46 "produced his greatest . . .": Robert Menzies, "Churchill at 75," *New York Times Magazine*, November 27, 1949, 37.

46 asked Churchill to lecture: F. L. McLuer to WSC, October 3, 1945, quoted in Gilbert, *Churchill*, vol. 8, 159.

46 "This is a . . .": Harry Truman to WSC, October 3, 1945, in G. W. Sand,

ed., *Defending the West: The Truman-Churchill Correspondence, 1945–1960* (Westport, CT: Praeger, 2004), 152.

46 Churchill's reply: WSC to Harry Truman, November 8, 1945, in Sand, ed., *Defending the West*, 153; John Ramsden, *Man of the Century: Winston Churchill and His Legend Since 1945* (London: HarperCollins, 2003), 159; Gilbert, *Churchill*, vol. 8, 172.

48 told a joint session: WSC, November 16, 1945, quoted in Gilbert, *Churchill*, vol. 8, 171.

49 motion of censure: WSC, November 27, 1945, in *New York Times*, November 28, 1945.

49 Conservative Central Council meeting: *New York Times*, November 29, 1945; *The Times*, November 29, 1945.

49 "Master Fighter": *The Times*, November 29, 1945.

49 "the gloomy vultures . . .": WSC, November 18, 1945, quoted in *New York Times*, November 29, 1945.

49 shot down Churchill's motion: *The Times*, December 7, 1945.

49 Clementine Churchill watched: Ibid.

50 Truman confirms his offer: Truman to WSC, November 16, 1945, in Sand, ed., *Defending the West*, 153.

50 press for a firm date: WSC to Truman, November 29, 1945, ibid., 154.

51 only because so many: the Dowager Duchess of Devonshire, author interview.

51 addressed the troops: Sir James Bisset, *Commodore: War, Peace and Big Ships* (Sydney: Angus & Robertson, 1961), 436.

52 "farewell to politics": Ibid.

V. The Wet Hen

53 "rather like a governess . . .": Anthony Eden to Cranborne, January 3, 1946, quoted in D. R. Thorpe, *Eden* (London: Chatto & Windus, 2003), 340.

54 struck many of the anti-appeasers: Lord Stockton, author interview.

54 barred Churchill from their meetings: Ibid.

54 "despised": Lord Carrington, author interview.

54 spoken of his intention to resign: Channon, *Chips*, October 3, 1940.

55 "losing his grip": Cadogan, *Diaries*, March 4, 1942.

55 "failing fast": Alanbrooke, *War Diaries*, May 1, 1944.

55 "Waiting to step . . .": Headlam, *Diaries*, July 9, 1948.

55 "an unworthy hope": Anthony Eden, diary entry, July 17, 1945, quoted in Eden, *The Reckoning*, 632.

55 out for ten years: Henderson, *Inside the Private Office*, 21.

55 "and get everything wrong": Oliver Harvey, *The War Diaries of Oliver Harvey, 1941–1945* (London: Collins, 1978), July 28, 1945.

56 "forever": Cranborne to Paul Emrys Evans, January 19, 1946, PEE Papers.

56 "evil day": Thorpe, *Eden*, 340.

56 "fed up with everything": Cranborne to Paul Emrys Evans, January 19, 1946, PEE Papers.

56 "anxious": Lascelles, *King's Counsellor*, January 8, 1946.
56 "the wet hen": The Dowager Duchess of Devonshire, author interview.
57 Baldwin warned: Thomas Jones, *A Diary with Letters, 1931–1950* (London: Oxford University Press, 1954), January 5–7, 1946.
57 "through with him": Quoted in David Dutton, *Anthony Eden* (London: Hodder, 1997), 232.
57 encouraged Eden: Cranborne to Anthony Eden, January 8, 1946, quoted ibid., 232–3.
57 "the backbone to . . .": Lord Carrington, author interview.
58 incident at UN dinner: Lascelles, *King's Counsellor*, January 10, 1946.
58 letter of rebuke: Ibid., January 11, 1946.
58 "made an ass": Ibid., January 14, 1946.
59 "strongly": Cranborne to Paul Emrys Evans, January 19, 1946, PEE Papers.
60 Cranborne was pleased: Cranborne to Paul Emrys Evans, January 19 and February 12, 1946, ibid.
60 Cranborne did not really believe: Cranborne to Paul Emrys Evans, February 12, 1946, ibid.

VI. Winnie, Winnie, Go Away

61 "delicious": Clementine Churchill to Mary Soames, January 18, 1946, quoted in Soames, *Clementine Churchill*, 441.
62 caught another cold: Clementine Churchill to Mary Soames, January 18, 1946, quoted ibid.; *New York Times*, January 20, 1946.
62 message from Truman: Gilbert, *Churchill*, vol. 8, 189.
62 Truman had had to cancel: *New York Times*, February 9, 1946.
62 rough flight to Washington: *New York Times*, February 11, 1946.
63 "an electric shock": The Earl of Halifax to Mr. Bevin, February 17, 1946, *BDFA,* part 4, series C, *North America*, vol. 1, *1946* (Bethesda, MD: University Publications of America, 1999), 64.
63 Stalin obsessively pored: Vladimir O. Pechatnov, "'The Allies Are Pressing on You to Break Your Will . . .': Foreign Policy Correspondence Between Stalin and Molotov and Other Politburo Members, September 1945–December 1946," trans. Vladislav M. Zubok (Working Paper No. 26, Cold War International History Project, Woodrow Wilson International Center for Scholars, September 1999), 9.
64 Churchill distressed by rumors re: Stalin's health: Diaries of William Lyon Mackenzie King, October 26, 1945, Library and Archives of Canada, Ottawa.
64 Harriman announced: *New York Times*, October 27, 1945.
64 Stalin breathed fire and fury: Pechatnov, "The Allies Are Pressing," 11.
64 "The appointment with . . .": Ibid., 13.
65 stay an additional night: *The Times*, February 12, 1946.
65 complained on the phone: Mackenzie King Diaries, February 28, 1946.
66 Soviet violation re: Iran compared to Rhineland: *New York Times*,

March 12, 1946; the Earl of Halifax to Mr. Bevin, March 2, 1946, *BDFA,* part 4, series C, vol. 1, 68.

66 "the Long Telegram": The Chargé in the Soviet Union (Kennan) to the Secretary of State, February 22, 1946, *FRUS: 1946*, vol. 6, *Eastern Europe, the Soviet Union* (1946).

66 would do nothing but good: WSC to Clement Attlee, March 7, 1946, quoted in Ramsden, *Man of the Century*, 178; Mackenzie King Diaries, March 25, 1946.

66 only if his presentation was well received: Mackenzie King Diaries, March 25, 1946.

67 "This is certainly not . . .": WSC, March 5, 1946, quoted in *New York Times*, March 6, 1946.

68 "notorious for his love . . .": The Earl of Halifax to Mr. Bevin, March 8, 1946, *BDFA*, part 4, series C, vol. 1, 25.

68 claimed not to have read the speech: Harry Truman, March 8, 1946, quoted in *New York Times*, March 9, 1946.

69 "a far more . . .": The Earl of Halifax to Mr. Bevin, March 23, 1946, *BDFA,* part 4, series C, vol. 1, 81.

69 "the sharpest jolt . . .": The Earl of Halifax to Mr. Bevin, March 10, 1946, ibid., 72.

69 "warmonger": Full text of Stalin interview in *New York Times*, March 14, 1946.

69 "the most violent . . .": Kennan to Byrnes, March 14, 1946, *FRUS: 1946*, vol. 6.

69 As Molotov later said: *Molotov Remembers*, 60.

70 scene at the Waldorf: The Earl of Halifax to Mr. Bevin, March 16, 1946, *BDFA*, part 4, series C, vol. 1, 77.

71 dizzy spells: Mackenzie King Diaries, March 25, 1946.

71 began to fall forward: Ibid.

VII. Imperious Caesar

72 intent that that night: Cranborne to Paul Emrys Evans, April 1 and April 5, 1946, PEE Papers.

72 precursors of a stroke: Mackenzie King Diaries, March 25, 1946.

73 to set off another war: Dutton, *Anthony Eden*, 320.

74 to keep the job from going to a younger rival: John Ramsden, *An Appetite for Power* (London: HarperCollins, 1999), 301.

75 "in an ever-increasing . . .": WSC to Anthony Eden, April 7, 1946, quoted in Gilbert, *Churchill*, vol. 8, 227.

75 "certainly one of . . .": Ibid.

75 "to play my part": Anthony Eden to WSC, April 10, 1946, ibid.

76 "happy unity": Cranborne to Paul Emrys Evans, March 14, 1946, PEE Papers.

76 "imperious Caesar": Cranborne to Salisbury, May 7, 1946, quoted in Simon

Ball, *The Guardsmen: Harold Macmillan, Three Friends, and the World They Made* (London: HarperCollins, 2004), 280.

77 "take their courage . . .": Cranborne to Jim Thomas, April 26, 1946, quoted ibid., 280.

77 caught up in party strife: Mackenzie King Diaries, May 30, 1946.

77 "Smuts and I . . .": Quoted in John Colville, *The Churchillians* (London: Weidenfeld & Nicolson, 1981), 175.

77 "I was apprehensive . . .": Winston S. Churchill, *The World Crisis* (London: Odhams Press, 1938), 57.

78 dinner party in honor of Mackenzie King: Mackenzie King Diaries, June 8, 1946.

78 urged him to devote himself: Ibid.

78 told Betty Cranborne: Cranborne to Paul Emrys Evans, June 9, 1946, PEE Papers.

79 to stand down: Cranborne to Anthony Eden, August 21, 1946, quoted in Ball, *The Guardsmen*, 280.

79 cared a good deal less: Lord Salisbury, author interview.

79 had the luxury: Hugh Cecil, author interview.

79 "Beware of rampant . . .": Channon, *Chips*, May 5, 1939.

79 distressed by the perception: Lord Carrington, author interview.

80 "rather as if . . .": Lord Stockton, author interview.

80 Zurich speech: WSC, September 19, 1946, in *New York Times*, September 20, 1946.

81 "Having possibly endangered . . .": Duff Cooper, *The Duff Cooper Diaries*, ed. John Julius Norwich (London: Weidenfeld & Nicolson, 2005), 420.

82 "to retire gracefully": WSC, October 5, 1946, in Robert Rhodes James, ed., *Churchill Speaks 1897–1963: Collected Speeches in Peace and War* (New York: Barnes & Noble Books, 1998), 896.

83 "rapidly losing ground": Cranborne to Paul Emrys Evans, January 9, 1947, PEE Papers.

83 "by some resolute . . .": Ibid.

84 "at all costs": Ibid.

84 "like Lord Chatham . . .": Ibid.

VIII. Plots and Plotters

85 fuel crisis: Violet Bonham Carter, *Daring to Hope: The Diaries and Letters of Violet Bonham Carter 1946–1969*, ed. Mark Pottle (London: Weidenfeld & Nicolson, 2000), February 11, 1947; Clementine Churchill to Mary Soames, February 16, 1947, quoted in Soames, *Clementine Churchill*, 456.

86 could no longer afford: Report on the Meeting of the State-War-Navy Coordinating Committee Subcommittee on Foreign Policy Information, February 28, 1947, *FRUS: 1947*, vol. 5, *The Near East and Africa* (1947).

86 March 12, 1947, censure vote: *Hansard*, March 12, 1947; *New York Times*, March 13, 1947.

86 Macmillan and Butler at Conservative meeting: *New York Times*, March 14, 1947.
87 beginning in late February: Ball, *The Guardsmen*, 282.
87 hernia operation: Moran, *Diaries*, June 1947.
87 gathering at Crookshank's: Stuart, *Within the Fringe*, 145.
87 disliked and distrusted: Lord Salisbury, author interview.
88 lost interest in Bobbety: Hugh Cecil, author interview.
88 affairs with other men: Lady Elizabeth Cavendish, the Dowager Duchess of Devonshire, Lady Lloyd, author interviews.
88 rekindled her interest: Hugh Cecil, author interview.
88 resented his lifelong affection: Lady Elizabeth Cavendish, the Dowager Duchess of Devonshire, author interviews.
88 doomed him: Lady Elizabeth Cavendish, the Dowager Duchess of Devonshire, author interviews.
88 Butler . . . threw in his lot: Headlam, *Diaries*, 487; Pierson Dixon, diary entry, July 21, 1947, quoted in Gilbert, *Churchill*, vol. 8, 341.
88 blocked by a number of factors: Sir Nicholas Henderson, author interview.
89 Stuart meets with Churchill: Stuart, *Within the Fringe*, 146–7; Thompson, *The Day Before Yesterday*, 87.
89 "Oh, you've joined . . .": Thompson, *The Day Before Yesterday*, 87.
90 "rude awakening": Pierson Dixon, diary entry, July 21, 1947, quoted in Gilbert, *Churchill*, vol. 8, 341.
90 1922 Committee appearance: Headlam, *Diaries*, July 31, 1947.
90 to forsake politics: Ibid., August 2, 1947.
90 "of nothing except . . .": Cooper, *Diaries*, November 24, 1947.
90 loathed Churchill: Ibid.
91 to lure and coax his pet birds: Cooper, *Autobiography*, 391.
91 "the most boring . . .": Lord Salisbury, author interview.
91 By Macmillan's own reckoning: Lord Stockton, author interview.
91 "political Siberia": Ibid.
92 always maneuvering, always watching: Lord Carrington, author interview.
92 making up for lost time: Lord Stockton, author interview.
92 "We need a Prime Minister . . .": *New York Times*, October 5, 1947.

IX. Before It Is Too Late

93 "in attendance": Lady Soames, author interview.
93 Churchill at wedding: Mackenzie King Diaries, November 20, 1947.
93 "We shall not . . .": Ibid.
94 Moscow decreed postwar world divided: *New York Times*, October 23, 1947.
94 Bevin told the Dominion leaders: Mackenzie King Diaries, November 24, 1947.
95 "If I were . . .": Quoted in Bonham Carter, *Daring to Hope*, August 12, 1948.
95 lunch party for Smuts and King: Mackenzie King Diaries, November 25, 1947.

95 "I am telling . . .": Ibid.
95 the whites of his bulging eyes: Ibid.
95 "The gleam in . . .": Ibid.
96 "admirers and hangers-on": Headlam, *Diaries*, March 7, 1949.
97 "always been too . . .": Ibid., February 23, 1949.
97 "Will there be . . .": WSC, January 23, 1948, quoted in *New York Times*, January 24, 1948.
98 "unless we are able to learn . . .": Editorial, *New York Times*, April 16, 1948.
98 serial sparked outrage: Channon, *Chips*, June 2, 1948.
99 Speaking at a mass rally: *New York Times*, June 29, 1948.
100 "something practical to . . .": WSC at Conservative Party Conference, Llandudno, October 9, 1948, quoted in *New York Times*, October 10, 1948.
100 "much heartburning": Salisbury to Paul Emrys Evans, October 4, 1948, PEE Papers.
100–101 "criminal error": Ibid.
101 delegates carrying the book: *New York Times*, October 6, 1948.
101 manifesto: Salisbury to Paul Emrys Evans, November 29, 1948, PEE Papers.
101 Kemsley makes public call: *New York Times*, October 19, 1948.
102 "devised with the . . .": Dwight D. Eisenhower, *Crusade in Europe* (Garden City: Doubleday, 1949), 396.
103 worst crisis of faith: Philip Goodhart with Ursula Branston, *The 1922: The Story of the Conservative Backbenchers' Parliamentary Committee* (London: Macmillan, 1973), 146.
104 "personal decks": Quoted ibid., 147.
104 fortunate he had not given the V-sign: Headlam, *Diaries*, March 3, 1949.
104 read the Liberal press: Lady Soames, author interview.
104 "great prestige": Quoted in Bonham Carter, *Champion Redoubtable*, August 1, 1944.
105 "gladly exchange the . . .": Mary Soames to Clementine Churchill, October 21, 1945, quoted in Soames, *Clementine Churchill*, 439.
105 "Very successful . . .": Channon, *Chips*, January 27, 1949.
106 "I do not mind . . .": Clementine Churchill to WSC, March 5, 1949, in Soames, ed., *Winston and Clementine.*
106 White House announced: *New York Times*, March 22, 1949.
107 "like a schoolboy": Macmillan, *Tides of Fortune*, 171.

X. The Dagger Is Pointed

108 "The dagger is . . .": Quoted in A. J. P. Taylor, *Beaverbrook* (London: Penguin, 1974), 122.
109 Moran's conclusion: Moran, *Diaries*, August 24, 1949.
110 rumors flew: Macmillan, *Tides of Fortune*, 179.
110 According to Sandys: Ibid.
110 In the days that followed: Moran, *Diaries*, 357.

110 "The dagger struck . . .": Quoted in Taylor, *Beaverbrook*, 758.
110 "notice to quit . . .": Moran, *Diaries*, October 11, 1951.
110 "wasted": Denis Kelly quoted in Gilbert, *Churchill*, vol. 8, 487.
110 worried press will notice: Moran, *Diaries*, 357.
111 attended the Lime Tree Stakes: *New York Times*, September 11, 1949.
112 "entranced": Channon, *Chips*, September 28, 1949.
112 "Over all there . . .": WSC, September 28, 1949, quoted in *New York Times*, September 29, 1949.
112 "like a soiled glove": *Hansard*, September 29, 1949.
113 "I hope, sir . . .": Quoted in Kay Halle, *The Irrepressible Churchill* (London: Robson, 1985), 246.
113 "I am prepared . . .": Ibid.
113 "I heard there . . .": *Time*, January 23, 1950.
114 "toil and moil": WSC telegram to Clementine Churchill, January 16, 1950, quoted in Gilbert, *Churchill*, vol. 8, 501.
114 "immense programme": WSC to Clementine Churchill, January 19, 1950, in Soames, ed., *Winston and Clementine*.
114 "When great events . . .": Quoted in Bonham Carter, *Daring to Hope*, January 16, 1950.
114 "It must be nearly . . .": Quoted in Lord Butler, *The Art of the Possible* (London: Hamish Hamilton, 1971), 152.
114 everything went misty: Moran, *Diaries*, January 24, 1950.
114 "I am not . . .": Quoted in Archibald Clark Kerr Diary, August 16, 1942.
115 Eden signaled the world: *New York Times*, February 14, 1950.
116 "Still, I cannot . . .": WSC, February 14, 1950, quoted in *New York Times*, February 15, 1950.
116 Labour reaction: *New York Times*, February 16, 1950.
117 "Of course, as I am reminded . . .": WSC, February 17, 1950, quoted in *New York Times*, February 18, 1950.
118 "the prelude to . . .": Macmillan, *Tides of Fortune*, 316.
118 inevitable: WSC to Sir Alan Lascelles, February 27, 1950, quoted in Gilbert, *Churchill*, vol. 8, 512.
118 Eden questioned the wisdom: Dutton, *Anthony Eden*, 236.

XI. Another Glass of Your Excellent Champagne

120 Beistegui Ball: Susan Mary Alsop, *To Marietta from Paris* (Garden City: Doubleday, 1965), 182–90; John Julius Norwich, *Trying to Please* (Wimborne Minster, Dorset: Dovecote Press, 2008), 140–2; Lanfranco Rasponi, *The International Nomads* (New York: Putnam's, 1966), 193–200; Charlotte Mosley, ed., *The Mitfords: Letters between Six Sisters* (New York: Harper, 2007), 272–4.
120 sink into the Grand Canal: Rasponi, *International Nomads*, 200.
121 "Anthony Eden nearly . . .": Halle, *The Irrepressible Churchill*, 255.
121 Seeking to counter sentiment: Harold Macmillan, *The Macmillan Diaries:*

Cabinet Years, 1950–1957, ed. Peter Catterall (London: Macmillan, 2003), June 11–14, 1951.

122 managed to extract: Malcolm Muggeridge, *Like It Was: The Diaries of Malcolm Muggeridge*, ed. John Bright-Holmes (London: Collins, 1981), 437.

122 whether he might safely swim: WSC to Dr Russell Brain, August 22, 1951, quoted in Gilbert, *Churchill*, vol. 8, 631.

122 Churchill in bathing trunks: *New York Times*, August 25, 1951.

123 "I found him . . .": Cooper, *Diaries*, August 1951.

123 "run interference": Lord Norwich, author interview.

123 watched his mother grow desperate: Ibid.

123 "I shall be . . .": Ibid.

124 "lacking in dignity": Cooper, *Diaries*, December 1951.

124 intended to lay down the burden: Macmillan, *Diaries*, October 4, 1951.

124 "physically able": Ibid., September 27, 1951.

124 "to younger hands": Ibid.

124 lost ground in recent months: Moran, *Diaries*, 361.

125 "at his best": Lady Soames, author interview.

125 Churchill confessed: Moran, *Diaries*, October 11, 1951.

125 "I do not hold . . .": WSC, October 8, 1951, in Winston S. Churchill, *Stemming the Tide: Speeches, 1951 and 1952*, ed. Randolph S. Churchill (Boston: Houghton Mifflin, 1954), 135.

126 "Do you think . . .": Quoted in Tony Benn, *Years of Hope: Diaries, Letters and Papers 1940–1962*, ed. Ruth Winstone (London: Hutchinson, 1994), October 31, 1951.

127 "cruel and ungrateful . . .": WSC, October 23, 1951, quoted in *New York Times*, October 24, 1951.

127 "It is the last . . .": Ibid.

127 "One may say . . .": Churchill, *Great Contemporaries*, 63.

XII. White with the Bones of Englishmen

129 "rather an ass": Salisbury to Paul Emrys Evans, October 29, 1951, PEE Papers.

130 infringement on the royal prerogative: Randolph Churchill, *The Rise and Fall of Sir Anthony Eden* (London: Macgibbon, 1959), 193–4.

130 "bullying people who . . .": P. J. Grigg quoted in Bonham Carter, *Daring to Hope*, January 3, 1957.

130 "into one throw": Lord Salisbury, author interview.

130 played havoc with his hearing: Moran, *Diaries*, November 9, 1951.

130 "a sea captain . . .": Lord Chandos quoted in Thompson, *The Day Before Yesterday*, 91.

130 die during the new Parliament: Channon, *Chips*, October 28, 1951.

130–131 in lieu of a hot water bottle: Rab Butler quoted in Thompson, *The Day Before Yesterday*, 91.

131 high words were exchanged: Evelyn Shuckburgh, *Descent to Suez: Foreign*

Office Diaries, 1951–1956, ed. John Charmley (New York: Norton, 1986), October 27, 1951.

132 Churchill intervened: Ibid.

132 "If he wants . . .": Quoted in Alistair Horne, *Macmillan, 1894–1956* (London: Macmillan, 1988), 341.

132 "in a most . . .": Macmillan, *Diaries*, October 28, 1951.

132 "either make or mar": Ibid.

133 poured out his anguish: Lady Elizabeth Cavendish, author interview.

133 far better campaigner: Lady Anne Tree, author interview.

133 extremely cautious: Lady Anne Tree, author interview.

134 "in great form": Macmillan, *Diaries*, October 30, 1951.

134 quoting Machiavelli: Cabinet Secretary's Notebooks, October 30, 1951, CAB 195/10, The National Archives, Public Record Office.

134 "up": Ibid.

134 "almost certain": Memorandum by the Director of the Office of British Commonwealth and Northern European Affairs (Raynor), October 18, 1951, *FRUS: 1951*, vol. 4, *Europe: political and economic developments (in two parts)*, part 1 (1951).

135 difficult for Truman to decline: Ibid.

135 "hasty public": Paper Prepared in the Department of State, n.d., *FRUS: 1951*, vol. 4, part 1; The Acting Secretary of State to the Embassy in the United Kingdom, October 25, 1951, ibid.

135 "Now that I am again": WSC to Stalin, November 5, 1951, Foreign Office 371/94841, quoted in Gilbert, *Churchill*, vol. 8, 659.

136 "cursed old age": Quoted in Simon Sebag Montefiore, *Stalin: The Court of the Red Tsar* (New York: Vintage, 2003), 614.

136 "finished": Quoted ibid.

136 "persecution mania": *Molotov Remembers*, 324.

136 Churchill reported to his Cabinet: Cabinet Secretary's Notebooks, November 12, 1951, CAB 195/10.

137 Eden welcomed and praised: *New York Times*, November 13, 1951, and November 14, 1951.

137 stomach spasms: Shuckburgh, *Descent to Suez*, November 4, 1951.

137 kill himself in the bargain: Salisbury to Paul Emrys Evans, October 29, 1951, PEE Papers.

138 traveled with a black tin box: Shuckburgh, *Descent to Suez*, 14.

138 heard through the wall: Ibid., November 21, 1951.

138 Eisenhower read advance extracts: Eisenhower to Hastings Ismay, October 2, 1951, in *The Papers of Dwight David Eisenhower*, ed. Louis Galambos and Daun Van Ee, vol. 12 (Baltimore: Johns Hopkins University Press, 1989), 589.

139 Eden breakfast with Eisenhower: Anthony Eden, *Full Circle* (Boston: Houghton Mifflin, 1960), 36; C. L. Sulzberger, *A Long Row of Candles: Memoirs and Diaries, 1934–1951* (New York: Macmillan, 1969), December 13, 1951.

139–140 liked the current European army plan: Sulzberger, *A Long Row of Candles*, December 13, 1951.

140 presented himself as: Dean Acheson to Truman, November 30, 1951, *FRUS: 1951*, vol. 3, *European security and the German question (in two parts)*, part 1 (1951); Memo of Conversation by Secretary of State with Eden, November 29, 1951, Rome, ibid.

140 a more cooperative mood: Sulzberger, *A Long Row of Candles*, December 13, 1951.

140 agreed not to attack: Gifford to Acheson, December 11, 1951, *FRUS: 1951*, vol. 3, part 1.

140 conferred with Adenauer: Ibid.

141 praising Shinwell: *The Times*, December 7, 1951.

141 hoped to ease French fears: Gifford to Acheson, December 19, 1951, *FRUS: 1951*, vol. 3, part 1.

141 "second string": Shuckburgh, *Descent to Suez*, December 16, 1951.

142 "very considerably": The Ambassador in France (Bruce) to the Secretary of State, December 19, 1951, *FRUS: 1951*, vol. 4, part 1.

142 could not last in office: Ibid.

142 "vigorously": Ibid.

142 expressed doubt: Ibid.

142 drinking whisky: Shuckburgh, *Descent to Suez*, December 16, 1951.

142 "deliberately grumpy": Sulzberger, *A Long Row of Candles*, December 13, 1951.

143 "to give Churchill hell": Ibid.

143 "Eisenhower blames Churchill . . .": Ibid.

143 "to put the bite . . .": Ibid.

143 "I want a . . .": The Ambassador in France (Bruce) to the Secretary of State, December 20, 1951, *FRUS: 1951*, vol. 3, part 1.

143 "pounded": Ibid.

143 "I am back . . .": Diary entry, December 21, 1951, in *The Papers of Dwight David Eisenhower*, vol. 12, 810.

144 "penniless beggar": The Ambassador in the United Kingdom (Gifford) to the Secretary of State, December 29, 1951, *FRUS: 1951*, vol. 4, part 1.

144 contacted the American ambassador: Ibid.

145 the captain came in: Shuckburgh, *Descent to Suez*, December 29, 1951.

146 "Your one value . . .": Lord Mountbatten, diary entry, December 31, 1951, quoted in Gilbert, *Churchill*, vol. 8, 673.

146 New Year's Eve: Shuckburgh, *Descent to Suez*, December 29, 1951; Moran, *Diaries*, January 1, 1952.

146 talked of his trepidation: Moran, *Diaries*, January 1, 1952.

147 on the presidential yacht: Memorandum by the Secretary of State of a Dinner Meeting Aboard the SS *Williamsburg* on the Evening of January 5, 1952, *FRUS: 1952–1954*, vol. 6, *Western Europe and Canada (in two parts)*, part 1 (1954); Notes by the Chairman of the Joint Chiefs of Staff (Bradley) of a Dinner Meeting Aboard the SS *Williamsburg* on the Evening of January 5, 1952, ibid.; Dean Acheson, *Present at the Creation* (New York: Norton, 1969), 597–600.

147 "definitely aging and . . .": The Ambassador of the United Kingdom (Gifford) to the Department of State, December 28, 1951, ibid.

147 "so that we . . .": Prime Minister Churchill to President Truman, December 10, 1951, *FRUS: 1952–1954*, vol. 6, part 1.

147 agenda and advisers: President Truman to Prime Minister Churchill, December 13, 1951, ibid.

148 "nervousness": Sir O. Franks, United States: Weekly Summary, Period 22nd–28th December 1951, December 29, 1951, *BDFA,* part 5, series C, vol. 2, *1952* (Bethesda, MD: LexisNexis, 2006), 206.

148 "by sheer weight . . .": Ibid.

148 he assured the president: Acheson, *Present at the Creation*, 597.

148 talk as equals: Moran, *Diaries*, January 5, 1952.

148 cut Churchill off: Shuckburgh, *Descent to Suez*, January 5, 1952.

148 "different circumstances": The Acting Secretary of State to the Embassy in the United Kingdom, December 9, 1951, *FRUS: 1952–1954*, vol. 6, part 1.

149 urged him not to concentrate: Moran, *Diaries*, January 7, 1952.

149 "with British blood": Memorandum by the Special Assistant to the Secretary of State (Battle) of a Meeting Between President Truman and Prime Minister Churchill, January 8, 1952, *FRUS: 1952–1954*, vol. 6, part 1.

150 "bravely and loyally": Shuckburgh, *Descent to Suez*, January 5, 1952.

150 tell night from day: Moran, *Diaries*, January 8, 1952.

150 distressed about press communiqué: Ibid.

151 conversation with Makins: Colville, *The Fringes of Power*, 600.

151 address to Congress: Sir O. Franks, United States: Weekly Summary, Period 12–18th January 1952, January 19, 1952, *BDFA,* part 5, series C, vol. 2, 215.

151 firm, strong voice: *The Times*, January 18, 1952.

151 visibly relieved: Moran, *Diaries*, January 17, 1952.

152 drafted a joint communiqué: Acheson, *Present at the Creation*, 601.

152 rip up the text: Ibid.

152 "Hurricane warnings . . .": Ibid.

152 Churchill during drafting of communiqué: US Secretary of Defense (Robert Lovett) to Eisenhower, January 24, 1952, *FRUS: 1952–1954*, vol. 6, part 1.

153 "majestic": Acheson, *Present at the Creation*, 602.

153 "upon that western . . .": Ibid.

154 raising the possibility of a conference: US Delegation Minutes of the Meeting of President Truman and Prime Minister Churchill, The White House, January 18, 1952, 3 p.m., *FRUS: 1952–1954*, vol. 6, part 1.

154 Truman easily brushed aside: Minutes, January 18, 1952, *FRUS: 1952–1954*, vol. 6, part 1.

154 Truman gave him several openings: Ibid.

154 Churchill sent Colville: Ibid.

155 "I accept": Secretary of Defense Lovett to the Supreme Allied Commander, Europe, Eisenhower, January 24, 1952, *FRUS: 1952–1954*, vol. 6, part 1.

XIII. Naked Among Mine Enemies

156 spent much of the train trip: Moran, *Diaries*, January 19, 1952.

156 intended to spend forty-eight hours: WSC to Clementine Churchill, January 20, 1952, in Soames, ed., *Winston and Clementine*.

156–157 received the archbishop and the mayor: *New York Times*, January 23, 1952.

157 dreamed that he was so sick: Moran, *Diaries*, January 26, 1952.

157 waiting at Southampton: *The Times*, January 29, 1952.

157 looked forward to: WSC to Clementine Churchill, January 20, 1952, in Soames, ed., *Winston and Clementine*.

158 his intention to reply: Cabinet Secretary's Notebooks, January 29, 1952, CAB 195/10.

158 "get away": Ibid.

158 "very anxious": Ibid.

158 "Let's see what . . .": Ibid.

158 "more convenient": *The Times*, January 30, 1952.

159 "tardy, timid and fatuous": *New York Times*, January 31, 1952.

159 "irritated": Richard Crossman, *The Backbench Diaries of Richard Crossman*, ed. Janet Morgan (New York: Holmes & Meier, 1981), February 5, 1952.

160 "I've got bad . . .": Quoted in Ben Pimlott, *The Queen: A Biography of Elizabeth II* (New York: John Wiley & Sons, 1997), 176.

160 tried to reassure him: Colville, *The Fringes of Power*, 601.

160 announcing that the King had died: Cabinet Secretary's Notebooks, February 6, 1952, CAB 195/10.

160 filled Churchill with dread: Ibid.

160 doubted that they would: Ibid.

160 Commons filled to overflowing: *The Times*, February 7, 1952.

161 "terribly disappointed": Crossman, *Backbench Diaries*, February 6, 1952.

161 "No doubt we . . .": Ibid.

161 "hard-boiled": Ibid., February 11, 1952.

161 "Directly you got . . .": Ibid.

161 "being much tired . . .": Macmillan, *Diaries*, February 7, 1952.

162 "In the end . . .": WSC, February 7, 1952, in Churchill, *Stemming the Tide*, 238.

163 in no rush: Cabinet Secretary's Notebooks, February 11, 1952, CAB 195/10.

163 "steadying effect": Quoted in Macmillan, *Diaries*, February 11, 1952.

163 a paternal attitude: Ibid., March 6, 1952.

163 "a real love-match": Lady Elizabeth Cavendish, author interview.

163 "I've seen the man . . .": Ibid.

163 "the power behind . . .": Macmillan, *Diaries*, February 8, 1952.

164 "Why was he . . .": Ibid.; Cabinet Secretary's Notebooks, February 8, 1952, CAB 195/10.

164 Mountbatten had been heard: Colville, *The Fringes of Power*, 602.

164 repeated the story: Cabinet Secretary's Notebooks, February 18, 1952, CAB 195/10.

165 top-secret information: Acheson, *Present at the Creation*, 618.

165 "naked among mine . . .": Ibid., 620.

165 "I can win . . .": Ibid., 621.

165 needed to focus: Cabinet Secretary's Notebooks, February 21, 1952, CAB 195/10.

165 Eden disclosed: Acheson, *Present at the Creation*, 618.

166 could not think of the words: Moran, *Diaries*, February 21, 1952.

XIV. I Live Here, Don't I?

167 recounted the telephone incident: Moran, *Diaries*, February 21, 1952.

167 her nightmare that someday: Anthony Montague Browne, *Long Sunset* (London: Cassell, 1995), 122.

168 her demeanor: Moran, *Diaries*, February 21, 1952.

168 meeting with Salisbury: Colville, *The Fringes of Power*, 603; Moran, *Diaries*, February 22, 1952.

169 Lascelles concurred: Moran, *Diaries*, February 22, 1952.

170 only his wife and doctor: Colville, *The Fringes of Power*, 603.

170 dull and listless: Moran, *Diaries*, February 29, 1952.

170 "cumbrous, ill-at-ease": Crossman, *Backbench Diaries*, February 26, 1952.

170 cheers that greeted him: *The Times*, February 27, 1952.

170 "He is looking . . .": Nigel Nicolson to Harold Nicolson, quoted in Gilbert, *Churchill*, vol. 8, 707.

170 Churchill's speech: *Hansard*, February 26, 1952.

171 "sat stunned and white": Crossman, *Backbench Diaries*, February 26, 1952.

172 took turns pummeling Butler: Cabinet Secretary's Notebooks, February 28, 1952, and March 4, 1952, CAB 195/10; Macmillan, *Diaries*, February 29, 1952.

172 taken the lead: Shuckburgh, *Descent to Suez*, 38.

172 Moran's letter: Moran, *Diaries*, March 12, 1952.

172 dismissed the letter: Ibid., March 13, 1952.

173 destroyed his medical records: Sebag Montefiore, *Stalin*, 202.

173 conferring with Cabinet about Stalin message: Cabinet Secretary's Notebooks, March 12, 1952, CAB 195/10.

173 Stalin proposed: *The Times*, March 11 and 12, 1952; Gifford to Dept of State, March 11, 1952, *FRUS: 1952–1954*, vol. 7, "Germany and Austria (in two parts)," part 1 (1954).

174 unnerved by heavy Conservative losses: *New York Times*, April 10, 1952; Macmillan, *Diaries*, April 8–10, 1952.

174 mocked and taunted: Channon, *Chips*, April 9, 1952.

174 assure the 1922 Committee: *New York Times*, April 10, 1952; Channon, *Chips*, April 9, 1952.

174 "old brain": Anthony Eden, diary entry, June 4, 1952, quoted in Rhodes James, *Anthony Eden*, 355.

174 met to formulate new ultimatum: Harry Crookshank, diary entry, June 16, 1952, quoted in Gilbert, *Churchill*, vol. 8, 736.
175 he was desperately hurt: Lady Soames, author interview.
175 incident in sitting rooom at Number Ten: Acheson, *Present at the Creation*, 662.
175 Eden suddenly fell ill: *New York Times*, June 30, 1952.
175 Foreign Office announced: *The Times*, July 1, 1952.
176 "fairly pulverized . . .": Channon, *Chips*, July 1, 1952.
176 flushed and furious: *New York Times*, July 31, 1952.
176 Colville sent for doctor: Moran, *Diaries*, June 30, 1952.
176 rather in love with her himself: The Dowager Duchess of Devonshire, author interview.
176 "latest admirer": Cooper, *Diaries*, November 24, 1947.
176 never stopped trying: Ibid.
176 "dynastic": Alan Clark, *The Tories* (London: Weidenfeld & Nicolson, 1998), 334.
177 "I thought it might . . .": WSC to Truman, August 22, 1952, *FRUS: 1952–1954*, vol. 10, *Iran* (1954).
177 "I do not . . .": Ibid.
177 Eden was furious: Colville, *The Fringes of Power*, August 22–25, 1952.
177 his best hope: Ibid., June 13–15, 1952.
178 "the best man": Sir O. Franks to Mr. Eden, 21st August 1952, *BDFA,* part 5, series C, vol. 2, 158.
178 "cheap politics": Sir O. Franks, United States Weekly Summary, 30th August–5th September 1952, September 6, 1952, ibid., 299.
178 "For your private . . .": Colville, *The Fringes of Power*, November 9, 1952.
178 "I send you . . .": WSC to Eisenhower, November 5, 1952, quoted in Gilbert, *Churchill*, vol. 8, 773; *New York Times*, November 6, 1952.
179 Churchill very cross: Shuckburgh, *Descent to Suez*, November 5, 1952.
179 "a sudden burst . . .": Channon, *Chips*, November 5, 1952.
179 did not even mention Churchill: Shuckburgh, *Descent to Suez*, November 20, 1952.
179 On the flight home: Ibid., November 26, 1952.
179 had been speaking to Clarissa: Ibid., November 28, 1952.
179 "begged": Ibid.
180 "rejuvenating" pills: *Time*, December 15, 1952.
180 unable to follow: Shuckburgh, *Descent to Suez*, December 2, 1952.
180 "pathetic": Ibid.
180 sounded out Lascelles: Ibid., December 4, 1952.
180 Eden told Churchill: Ibid., December 8, 1952.
181 "I am delighted . . .": Eisenhower to WSC, December 17, 1952, in *The Papers of Dwight David Eisenhower*, vol. 13 (Baltimore: Johns Hopkins University Press, 1989), 1450.
182 best work was still ahead: Crossman, *Backbench Diaries*, February 5, 1952.
182 "completely fatuous": Diary entry, January 6, 1953, in *The Papers of Dwight David Eisenhower*, vol. 13, 1481.

182 "would turn over . . .": Ibid.

182 told an audience in Detroit: *New York Times*, June 15, 1952.

182 "no political directive": Dwight Eisenhower to Forrest Carlisle Pogue, February 20, 1952, in *The Papers of Dwight David Eisenhower*, vol. 13, 1001.

182 no role in political decisions: Dwight Eisenhower to Fred Kent, April 17, 1952, and Dwight Eisenhower to Richard Nixon, October 1, 1952, ibid., 1183.

183 "much vigor": Prime Minister to Secretary of State and Chancellor of the Exchequer, January 8, 1953, PREM 11/422.

183 "quite welcome": Ibid.

184 "certain power": Ibid.

184 "in too great . . .": Ibid.

XV. If Nothing Can Be Arranged

185 "a real man . . .": Colville, *The Fringes of Power*, January 11, 1953.

185 seemed to listen: Prime Minister to Secretary of State and Chancellor of the Exchequer, January 8, 1953, PREM 11/422.

185 rejected Churchill's suggestion: Colville, *The Fringes of Power*, January 7, 1953.

186 "an increasing liability": Anthony Eden, diary entry, January 23, 1953, quoted in Rhodes James, *Anthony Eden*, 358.

186 Lascelles assured: Shuckburgh, *Descent to Suez*, January 19, 1953.

186 tempted Macmillan: Macmillan, *Diaries*, January 17, 1953.

186 Walter Gifford: Diary entry, February 13, 1953, in *The Papers of Dwight David Eisenhower*, vol. 14 (Baltimore: Johns Hopkins University Press, 1996), 43.

187 For some twelve hours: Sebag Montefiore, *Stalin*, 640.

187 "killer doctors": *Pravda*, January 13, 1953.

187 To allay public fears: Sir A. Gascoigne to Foreign Office, March 4, 1953, PREM 11/540.

187 "contorted": Quoted in Sebag Montefiore, *Stalin*, 642.

187 joint leadership: Sir A. Gascoigne to Foreign Office, March 7, 1953, PREM 11/540.

187 paunchy, unkempt: Sir A. Gascoigne to Mr. Eden, March 19, 1953, *BDFA*, part 5, series A, *Soviet Union and Finland*, vol. 3, *1953* (Bethesda, MD: Lexis Nexis, 2007), 52.

187 compared to Machiavelli: Memorandum of Discussion at the 150th Meeting of the National Security Council, June 18, 1953, *FRUS: 1952–1954*, vol. 7, part 2.

187 to come out on top: Sir A. Gascoigne to Foreign Office, March 4, 1953, PREM 11/540.

187 failed to accord: P. Mason to Foreign Office, March 7, 1953, PREM 11/540.

187 "like hawks": Ibid.

188 "I remember our . . .": Prime Minister to President Eisenhower, March 10, 1953, PREM 11/422.

189 to throw Eden together with Molotov: Prime Minister to Secretary of State for Foreign Affairs, March 28, 1953, PREM 11/422.

189 "interim objective": Ibid.

189 "very keen": Shuckburgh, *Descent to Suez*, April 1, 1953.

190 jailed doctors had been released: The Chargé in the Soviet Union (Beam) to the Department of State, April 4, 1953, *FRUS: 1952–1954*, vol. 8, *Eastern Europe; Soviet Union; Eastern Mediterranean* (1954).

190 the most dangerous moment: Prime Minister Churchill to President Eisenhower, April 11, 1953, *FRUS: 1952–1954*, vol. 6, part 1.

190 distressed that he had missed: Memorandum of Telephone Conversation with the President by the Secretary of State, March 16, 1953, *FRUS: 1952–1954*, vol. 8.

191 "We do not know . . .": Prime Minister Churchill to President Eisenhower, April 11, 1953, *FRUS: 1952–1954*, vol. 6, part 1.

191 claiming that had Stalin: Memorandum of National Security Council Discussion, March 11, 1953, *FRUS: 1952–1954*, vol. 8.

192 "Well, maybe Churchill's . . .": Emmet John Hughes, *The Ordeal of Power* (New York: Atheneum, 1963), 113.

192 "prayerfully": President Eisenhower to Prime Minister Churchill, April 11, 1953, *FRUS: 1952–1954*, vol. 6, part 1.

193 Eden's operation: Thorpe, *Eden*, 384–5; Rhodes James, *Anthony Eden*, 362–3; David Owen, *In Sickness and in Power* (Westport, CT: Praeger, 2008), 109–10.

193 "slipped": Rhodes James, *Anthony Eden*, 362.

193 agreed to tone down: President Eisenhower to Prime Minister Churchill, April 11, 1953, *FRUS: 1952–1954*, vol. 6, part 1.

193 "massive and magnificent": quoted in *New York Times*, April 18, 1953.

194 "In my opinion . . .": Prime Minister Churchill to President Eisenhower, April 21, 1953, PREM 11/422.

194 "If nothing can be arranged": Prime Minister Churchill to President Eisenhower, April 21, 1953, PREM 11/422.

194 Lord Swinton: Moran, *Diaries*, April 24, 1953.

195 "extraordinarily well . . .": Bonham Carter, *Daring to Hope*, April 15, 1953.

195 Moran warned: Moran, *Diaries*, April 24, 1953.

195 spoke of not rushing: President Eisenhower to Prime Minister Churchill, April 25, 1953, PREM 11/422.

196 "who could take up . . .": CC (53) 29th Conclusions, Minute I, PREM 11/422.

196 "hadn't started this . . .": Cabinet Secretary's Notebooks, April 28, 1953, CAB 195/11.

196 sent Eisenhower his draft: Prime Minister Churchill to President Eisenhower, May 4, 1953, *FRUS: 1952–1954*, vol. 8.

197 advised him not to go: President Eisenhower to Prime Minister Churchill, May 5, 1953, ibid.

197 "It is the . . .": *New York Times*, May 12, 1953.

198 "the diplomatic leadership . . .": *New York Times*, May 18, 1953.

198 "faculties and judgment": Memorandum of Discussion at the 145th Meeting of the National Security Council, May 20, 1953, *FRUS: 1952–1954*, vol. 15, *Korea (in two parts)*, part 1.

198 "Sir Winston Churchill . . .": *Hansard*, July 29, 1953.

199 pretended not to hear: Eisenhower phone call to Churchill, May 20, 1953, *FRUS: 1952–1954*, vol. 6, part 1; Cabinet Secretary's Notebooks, May 21, 1953, CAB 195/11.

199 "in a mood . . .": Macmillan, *Diaries*, May 21, 1953.

200 "frankly skeptical": Lord Salisbury to Prime Minister Churchill, June 11, 1953, PREM 11/428.

200 "a quiet time . . .": Lord Salisbury, *Hansard*, July 29, 1953.

200 "Eisenhower has followed . . .": Cabinet Secretary's Notebooks, May 21, 1953, CAB 195/11.

200 "a direct result": Lord Salisbury to Prime Minister Churchill, June 11, 1953, PREM 11/428.

200 "completely negative attitude": Ibid.

200 "Yes, sir, it . . .": *New York Times*, May 22, 1953.

XVI. The Abdication of Diocletian

201 the detective suggested: Edmund Murray, "The Churchill I Knew" (paper presented at the Churchill Centre, Surrey, England, 9th International Conference, June 13, 1992).

201 encounter with Bevan: Ibid.

202 leaned on the Soviets: WSC to Vyacheslav Molotov, June 2, 1953, PREM 11/420.

203 "big game": Macmillan, *Diaries*, June 1–7, 1953.

203 "would not trust . . .": Ibid., September 1, 1953.

203 "through": Ibid.

203 Churchill late: Colville, *The Fringes of Power*, 666.

203 the race: *The Times*, June 8, 1953; *Time*, June 15, 1953.

203 204 "a most serious . . .": Macmillan, *Diaries*, June 1–7, 1953.

204 "an understanding attitude": Lord Salisbury to Prime Minister Churchill, June 11, 1953, PREM 11/428.

204 "great temptation": Ibid.

204 "I am sure . . .": Ibid.

204 agreed to a firm set of dates: WSC to Eisenhower, June 20, 1953, in Peter G. Boyle, ed., *The Churchill-Eisenhower Correspondence, 1953–1955* (Chapel Hill: University of North Carolina Press, 1990), 78.

204 held forth with brio: Colville, *The Fringes of Power*, 625.

204 "I want the . . .": Quoted in Kenneth Clark, *The Other Half* (New York: Harper, 1977), 128.

205 leaning heavily: Colville, *The Fringes of Power*, 626.

205 spasm of a small artery: Moran, *Diaries*, June 24, 1953.

206 "Harold, you might . . .": Macmillan, *Tides of Fortune*, 516.

206 He wanted to attend: Moran, *Diaries*, June 25, 1953.

207 walk unaided: Lady Soames, author interview.

207 no one was to be told: Colville, *The Fringes of Power*, 626.

207 Had he left London: Lady Soames, author interview.

207 In defiance: Colville, *The Fringes of Power*, 626.

208 thrombosis spreading: Moran, *Diaries*, June 26, 1953.

209 Salisbury and Butler visit: Butler, *The Art of the Possible*, 169–70; Moran, *Diaries*, June 26, 1953; Macmillan, *Diaries*, July 4, 1953; Cabinet Secretary's Notebooks, June 29, 1953, CAB 195/11.

213 live through the weekend: Soames, *Clementine Churchill*, 475.

213 Beaverbrook visit: Moran, *Diaries*, June 28, 1953.

213 Cabinet meeting: Cabinet Secretary's Notebooks, June 29, 1953, CAB 195/11; Macmillan, *Diaries*, July 4, 1953.

214 "intermediate": Cabinet Secretary's Notebooks, June 29, 1953, CAB 195/11.

214 "As he had . . .": John Wheeler-Bennett, ed., *Action This Day* (New York: St. Martin's Press, 1969), 44.

214 "Four years ago . . .": WSC to Eisenhower, July 1, 1953, in Boyle, ed., *The Churchill-Eisenhower Correspondence*, 82.

215 "I have had . . .": Macmillan, *Diaries*, July 2, 1953.

215 two key goals: WSC, "Policy Towards the Soviet Union and Germany," July 6, 1953, PREM 11/420.

215 read the prime minister's statement: Cabinet Secretary's Notebooks, July 6, 1953, CAB 195/11.

216 news from Moscow: First tripartite Foreign Ministers Meeting, July 19, 1953, *FRUS: 1952–1954*, vol. 5, *Western European Security (in two parts)*, part 2.

217 He nimbly argued: Opening Public Statements at First Tripartite Meeting, July 10, 1953, PREM 11/425.

217 "an unknown quantity": Ibid.

217 "when practicable": First Tripartite Foreign Ministers Meeting, July 10, 1953, *FRUS: 1952–1954*, vol. 5, part 2.

217 cabled London for instructions: Message from Lord Salisbury to the Chancellor of the Exchequer, July 11, 1953, PREM 11/425.

218 "They've bitched . . .": Moran, *Diaries*, July 14, 1953.

218 "Meanwhile, I am . . .": Prime Minister Churchill to President Eisenhower, July 17, 1953, *FRUS: 1952–1954*, vol. 6, part 1.

218 great hopes: *Hansard*, July 21, 1953.

219 "the lesser men . . .": Ibid.

219 impatient to hear: Moran, *Diaries*, July 21, 1953.

219 "I like to . . .": President Eisenhower to Prime Minister Churchill, July 20, 1953, *FRUS: 1952–1954*, vol. 6, part 1.

220 "violently Russophobe": Colville, *The Fringes of Power*, July 31–August 4, 1953.

220 "by every possible . . .": Meeting Between President Eisenhower and Leaders of the Three Delegations, July 11, 1953, PREM 11/425.

221 "triumphal": Shuckburgh, *Descent to Suez*, July 21, 1953.
222 would not like to be asked: Ibid., July 23, 1953.
222 "two invalids": Colville, *The Fringes of Power*, July 31–August 4, 1953.
223 "What! Can't you . . .": Clarissa Eden, diary entry, July 27, 1953, quoted in Clarissa Eden, *Clarissa Eden: A Memoir—From Churchill to Eden*, ed. Cate Haste (London: Weidenfeld & Nicolson, 2007), 142.
223 "Anthony must get . . .": Ibid.
223 vowed beforehand: Colville, *The Fringes of Power*, July 31–August 4, 1953.
223 to beard Churchill: Shuckburgh, *Descent to Suez*, July 28, 1953.
223 "I have been . . .": Ibid.
224 "I noted with . . .": *Hansard*, July 29, 1953.
224 conversation with Aldrich: The Ambassador in the United Kingdom (Aldrich) to the Department of State, September 11, 1953, *FRUS: 1952–1954*, vol. 6, part 1.
225 racing to orchestrate: Colville, *The Fringes of Power*, July 31–August 4, 1953.
225 "He is a . . .": Clarissa Eden, diary entry, July 27, 1953, quoted in Eden, *Clarissa Eden*, 142.
225 had better sleep on it: Ibid.
226 Violet Bonham Carter visited: Bonham Carter, *Daring to Hope*, August 6, 1953.
227 recites Gibbon: Moran, *Diaries*, August 8, 1953.
227 "more charming and . . .": Macmillan, *Diaries*, August 10, 1953.
227 Clementine sure Eisenhower does not want to see WSC: Moran, *Diaries*, August 16, 1953.
228 presided over his first Cabinet: Cabinet Secretary's Notebooks, August 18, 1953, CAB 195/11.
228 "dazed and gray": Soames, *Clementine Churchill*, 477.
228 wondered aloud: Moran, *Diaries*, August 19, 1953.
229 quarreled bitterly: Soames, *Clementine Churchill*, 478.
229 on an ebb tide: Ibid., 479.
230 "fit and ready . . .": *The Times*, October 2, 1953.
230 look as if he were not fit: Shuckburgh, *Descent to Suez*, October 1, 1953.
230 Churchill's dinner: Anthony Eden, diary entry, October 1, 1953, quoted in Rhodes James, *Anthony Eden*, 371; Clarissa Eden, diary entry, October 2, 1953, quoted in Eden, *Clarissa Eden*, 147; Shuckburgh, *Descent to Suez*, October 2, 1953.
231 initially saw Eisenhower: Shuckburgh, *Descent to Suez*, October 5, 1953.
232 sent off a message: Prime Minister Churchill to President Eisenhower, October 7, 1953, in Boyle, ed., *The Churchill-Eisenhower Correspondence*, 89.
232 Churchill rehearses: Moran, *Diaries*, October 9, 1953.
233 Eisenhower icily refused: President Eisenhower to Prime Minister Churchill, October 8, 1953, in *The Papers of Dwight David Eisenhower*, vol. 14, 458.
233 on the spot: Shuckburgh, *Descent to Suez*, October 9, 1953.
233 "I haven't got . . .": Ibid.
234 "What subjects are . . .": Colville, *The Fringes of Power*, October 1953.

234 "I am very . . .": Prime Minister Churchill to President Eisenhower, October 9, 1953, in Boyle, ed., *The Churchill-Eisenhower Correspondence*, 91.

234 "whether he would . . .": Edward Heath, *The Course of My Life* (London: Hodder, 1998), 158.

235 "If I stay . . .": WSC, October 10, 1953, quoted in *New York Times*, October 11, 1953.

235 "triumphant return to . . .": *The Times*, October 12, 1953.

236 "would be walking a tightrope . . .": Memo of Conversation, October 14, 1953, *FRUS: 1952–1954*, vol. 7, part 1.

236 dinner party for Dulles at Number Ten: The Secretary of State to the President, October 16, 1953, ibid.

237 "loyally": Ibid.

237 farewell luncheon: The Ambassador in the United Kingdom (Aldrich) to the Department of State, October 19, 1953, ibid.

239 whether he wished to respond: President Eisenhower to Arthur Hays Sulzberger, September 11, 1953, in *The Papers of Dwight David Eisenhower*, vol. 14, 511.

240 "The Soviet answer . . .": Prime Minister Churchill to President Eisenhower, November 5, 1953, PREM 11/418.

240 must not be billed: President Eisenhower to Prime Minister Churchill, November 7, 1953, PREM 11/418.

240 "I shall go . . .": Halle, *The Irrepressible Churchill*, 274.

XVII. I Have a Right to Be Heard!

241 "Ike": Moran, *Diaries*, December 3, 1953.

242 to douse press expectations: Sulzberger, *A Long Row of Candles*, November 23, 1953.

242 forced: *New York Times*, December 5, 1953.

242 "trotting behind . . .": Shuckburgh, *Descent to Suez*, December 4, 1953.

242 "pirated": Colville, *The Fringes of Power*, December 4, 1953.

242 broached the possibility: Diary, December 4, 1953, in *The Papers of Dwight David Eisenhower*, vol. 15 (Baltimore: Johns Hopkins University Press, 1996), 729.

242 convened that evening: Minutes of First Plenary Meeting, December 4, 1953, PREM 11/418.

243 laid out the strategy: Policy of Soviet Union, December 4, 1953, PREM 11/668.

243 "vulgar": Shuckburgh, *Descent to Suez*, December 4, 1953.

243 "very violent": Colville, *The Fringes of Power*, December 4, 1953.

243 a woman of the streets: Ibid.

243 "pained looks": Ibid.

244 "I don't know . . .": Ibid.

244 squared off on atomic policy: Bermuda, December 5, 1953, Meeting Held 11:30, Notes Written by Admiral Strauss, Dwight David Eisenhower Library.

245 "I have been . . .": Moran, *Diaries*, December 7, 1953.
245 "kind": Shuckburgh, *Descent to Suez*, January 18 and 19, 1954.
246 "gaga": Ibid., March 31, 1954.
246 Soames concluded: Moran, *Diaries*, February 7, 1954.
246 "atheistic materialism": President Eisenhower to Prime Minister Churchill, February 9, 1954, *FRUS: 1952–1954*, vol. 6, part 1.
246 "for the struggle": Ibid.
246 even more alarmed: Shuckburgh, *Descent to Suez*, March 1, 1954.
246 "start out": Memorandum of a Meeting of President Eisenhower and Prime Minister Churchill at the White House, June 25, 1954, *FRUS: 1952–1954*, vol. 6, part 1.
247 He believed that: Prime Minister Churchill to President Eisenhower, March 9, 1954, ibid.
247 cryptic message: Christopher Soames to the Prime Minister, March 17, 1954, PREM 11/668.
247 Eisenhower checked in: President Eisenhower to Prime Minister Churchill, March 19, 1954, *FRUS: 1952–1954*, vol. 6, part 1.
248 made things worse: Press conference, March 24, 1954, in *New York Times*, March 25, 1954.
249 "more statesmanlike": Julian Amery, *Approach March* (London: Hutchinson, 1973), 434.
249 Attlee spoke first: *Hansard*, April 5, 1954.
249 Churchill's performance: *The Times*, April 6, 1954; *New York Times*, April 6, 1954; *Hansard*, April 5, 1954.
250 "I have a right to be heard!": *New York Times*, April 6, 1954.
250 Macmillan guessed: Macmillan, *Diaries*, April 6, 1954.
250 White-faced: Montague Browne, *Long Sunset*, 180.
250 swayed slightly: Crossman, *Backbench Diaries*, April 6, 1954.
250 Woolton relieved: Moran, *Diaries*, April 5, 1954.
250 The doctor worried: Ibid.
250 Boothby later claimed: *The Times*, April 6, 1954.
250 Boothby later claimed: Robert Rhodes James, *Robert Boothby* (New York: Viking, 1991), 373.
251 Two ministers: Montague Browne, *Long Sunset*, 180.
251 no later than Whitsun: Shuckburgh, *Descent to Suez*, April 6, 1954.
251 must not under any circumstances: Ibid.
252 "warrant": WSC to Bernard Baruch, August 29, 1954, PREM 11/669.
252 Eisenhower asked Churchill: President Eisenhower to Prime Minister Churchill, April 4, 1954, in *The Papers of Dwight David Eisenhower*, vol. 15, 1002; Cabinet Secretary's Notebooks, April 7, 1954, CAB 195/12; Eden Memo to Cabinet, April 7, 1954, CAB 129/67 post war memoranda.
253 "air action": Cabinet Secretary's Notebooks, April 7, 1954, CAB 195/12.
253 "very much like . . .": Prime Minister Churchill to President Eisenhower, April 22, 1954, in Boyle, ed., *The Churchill-Eisenhower Correspondence*, 139.

254 Eisenhower naturally assumed: Eisenhower to Dulles, April 23, 1954, *FRUS: 1952–1954*, vol. 13, *Indochina (in two parts)*, part 1.

254 Churchill wrote to inform Eden: WSC to Anthony Eden, May 12, 1954, PREM 11/291.

254 told the Cabinet: Cabinet Secretary's Notebooks, June 5, 1954, CAB 195/12.

254 "precipitate action": Cabinet Secretary's Notebooks, May 24, 1954, CAB 195/12.

254 "prize noodle": Salisbury to Anthony Eden, May 9, 1954, quoted in John Charmley, *Churchill's Grand Alliance* (New York: Harcourt, 1995), 284.

254 "a strong deterrent . . .": Cabinet Secretary's Notebooks, May 24, 1954, CAB 195/12.

254 to counter any arguments: Cabinet Secretary's Notebooks, June 5, 1954, CAB 195/12.

255 "a very perfect . . .": WSC to Clementine Churchill, June 5, 1954, in Soames, ed., *Winston and Clementine*.

255 asked Churchill to step aside: Anthony Eden to WSC, June 7, 1954, quoted in Gilbert, *Churchill*, vol. 8, 989.

255 "see what emerges . . .": WSC to Anthony Eden, June 11, 1954, quoted ibid., 989–90.

255 "terrified": Quoted in John Young, *Winston Churchill's Last Campaign* (Oxford: Oxford University Press, 1996), 266.

255 "like squids or . . .": Quoted ibid.

256 Eden and Dulles: Memorandum of Meeting of Secretary of State Dulles and Foreign Secretary Eden, June 25, 1954, *FRUS: 1952–1954*, vol. 6, part 1.

256 three-quarters of an hour: Editorial Note, no. 465, ibid.

256 Eisenhower's answer: Colville, *The Fringes of Power*, June 25, 1954.

257 put his proposal in writing: Memorandum of a Meeting of President Eisenhower and Prime Minister Churchill at the White House, June 25, 1954, 3 p.m., *FRUS: 1952–1954*, vol. 6, part 1.

257 until he was back in Britain: Memorandum of a Meeting of President Eisenhower and Prime Minister Churchill at the White House, June 26, 1954, 11 a.m., ibid.

257 "reconnaissance in force": Memorandum of a Meeting of President Eisenhower and Prime Minister Churchill at the White House, June 25, 1954, 3 p.m., ibid.

257 "You are to . . .": Macmillan, *Diaries*, July 6, 1954.

258 Dulles expressed doubt: Memorandum of a Conversation by the Secretary of State, June 27, 1954, *FRUS: 1952–1954*, vol. 6, part 1.

258 "very carefully weighed . . .": Ibid.

258 reported the conversation: Ibid.

258 rooting for Butler: Memorandum of a Conference at the White House, May 5, 1954, *FRUS: 1952–1954*, vol. 13, part 2.

258 in the hope of extracting: Colville, *The Fringes of Power*, July 2, 1954.

XVIII. An Obstinate Pig

260 incident in the grill room: Moran, *Diaries*, July 1, 1954.

260 "bashful": Colville, *The Fringes of Power*, July 2, 1954.

261 raise objections: Ibid., July 16, 1954.

261 offered to personally deliver: Ibid., July 2, 1954.

262 out the porthole: Clarissa Eden, diary entry, July 8, 1954, quoted in Eden, *Clarissa Eden*, 164.

262 "in principle": Colville, *The Fringes of Power*, July 2, 1954.

262 "I propose to . . .": From Secretary of State on Board RMS *Queen Elizabeth* to Foreign Office, July 2, 1954, PREM 11/669.

262 Eden did not: Cabinet Secretary's Notebooks, July 8, 1954, CAB 195/12.

262 Eden pouted: Colville, *The Fringes of Power*, July 3, 1954.

263 Churchill grumbled: Ibid.

263 "Presume my message . . .": From RMS *Queen Elizabeth* to Foreign Office, July 3, 1954, PREM 11/669.

263 Butler in Norfolk: Chancellor of the Exchequer to Prime Minister Churchill, July 21, 1954, PREM 11/669.

263 suggestions for minor edits: From Foreign Office to RMS *Queen Elizabeth*, July 3, 1954, PREM 11/669.

263 Hayter, went to the Kremlin: From Moscow to Foreign Office, July 4, 1954, PREM 11/669.

263–264 "Your idea about . . .": To the Prime Minister, Mr. W. Churchill, From V. M. Molotov, July 5, 1954, PREM 11/669.

264 summoned Salisbury and Macmillan: Macmillan, *Diaries*, July 6, 1954.

265 "In the light . . .": Prime Minister Churchill to President Eisenhower, July 7, 1954, *FRUS: 1952–1954*, vol. 6, part 1.

265 messages from Foreign Office colleagues: Shuckburgh, *Descent to Suez*, July 6, 1954.

266 Cabinet meeting: Cabinet Secretary's Notebooks, July 7, 1954, CAB 195/12.

266 "bombshell": Macmillan, *Diaries*, July 7, 1954.

266 "blithely unconscious": Ibid.

266 "must, however, inform . . .": Cabinet Secretary's Notebooks, July 7, 1954, CAB 195/12.

266 "tacitly": Ibid.

266 "blank surprise": Macmillan, *Diaries*, July 7, 1954.

266 "disgust": Ibid.

267 Eden spoke: Cabinet Secretary's Notebooks, July 7, 1954, CAB 195/12.

267 Salisbury put Churchill on notice: Ibid.

267 "temper": Ibid.

267 "critical or hostile": Ibid.

267 "The United States can't veto . . .": Ibid.

268 cavalier: Macmillan, *Diaries*, July 7, 1954.

268 conversation at Ismay dinner: Ibid.

269 Eisenhower regarded it as proof: Notes of Telephone Conversation between

Eisenhower and Dulles, July 7, 1954, *The Papers of Dwight David Eisenhower*, vol. 15, 1167.

269 "only another stroke": Ibid.

269 "went fishing": Sir Roger Makins to Anthony Eden, July 7, 1954, PREM 11/669.

269 "could not prevent . . .": Ibid.

269 "You did not . . .": President Eisenhower to Prime Minister Churchill, July 7, 1954, PREM 11/669.

270 "I hope you . . .": Prime Minister Churchill to President Eisenhower, July 8, 1954, PREM 11/669.

271 "better": Cabinet Secretary's Notebooks, July 8, 1954, CAB 195/12.

271 paint himself as blameless: Ibid.

271 stakes so high: Colville, *The Fringes of Power*, July 16, 1954.

271 "absolute right": Cabinet Secretary's Notebooks, July 8, 1954, CAB 195/12.

271 remedy of ministers: Ibid.

272 Action, he conceded: Ibid.

272 "He could have . . .": Ibid.

272 "in mind": Ibid.

272 "Only two days . . .": Ibid.

272 "What was the . . .": Ibid.

272 "I may have . . .": Ibid.

273 "If I found . . .": Ibid.

273 "Of course I . . .": President Eisenhower to Prime Minister Churchill, July 9, 1954, PREM 11/669.

273 "I am very . . .": Prime Minister Churchill to President Eisenhower, July 9, 1954, PREM 11/669.

273 "I am unalterably . . .": Cabinet Secretary's Notebooks, July 9, 1954, CAB 195/12.

274 "Other members of . . .": Ibid.

274 "dead white": Macmillan, *Diaries*, July 9, 1954.

274 "puce": Ibid.

274 "I should greatly . . .": Ibid.

274 visit to Hatfield: Ibid.

275 thought he would cite: Shuckburgh, *Descent to Suez*, July 12, 1954.

275 "You, of course . . .": President Eisenhower to Prime Minister Churchill, July 13, 1954, PREM 11/669.

275 "minimize": Ibid.

275 "glum and worried": Macmillan, *Diaries*, July 13, 1854.

275 "give a damn": Colville, *The Fringes of Power*, July 16, 1954.

276 "Yes, I do . . .": Clementine Churchill to John Colville, quoted in Gilbert, *Churchill*, vol. 8, 1038.

276 all the cards: Macmillan, *Diaries*, July 16, 1954.

277 constitutional niceties: Ibid.

277 resignation would be foolish: Ibid.

277 such a fuss: Colville, *The Fringes of Power*, July 16, 1954.

277 Only now did Churchill: Ibid.
278 "this stickler for . . .": For Secretary of State from Prime Minister, July 17, 1954, PREM 11/669.
278 "compound the sins . . .": Ibid.
278 disputed him in writing: Chancellor of the Exchequer to Prime Minister Churchill, July 21, 1954, PREM 11/669.
278 "gravely concerned": Memorandum by the Under Secretary of State (Smith) to the Secretary of State, July 26, 1954, *FRUS: 1952–1954*, vol. 6, part 1.
279 "I am an obstinate pig": WSC to Eisenhower, August 8, 1954, ibid.
279 expected Eden's support: Minute to Prime Minister, July 22, 1954, PREM 11/669.
279 went home without her: Clarissa Eden, diary entry, July 22, 1954, in Eden, *Clarissa Eden*, 168.
279 in the middle of the night: Ibid.
279 on reflection he would not: Minute to Prime Minister, July 22, 1954, PREM 11/669.
279 "white and tense": Macmillan, *Diaries*, July 23, 1954.
279 "no message need . . .": Cabinet Secretary's Notebooks, July 23, 1954, CAB 195/12.
280 "On reflection,": Ibid.
280 "During this period . . .": Ibid.
281 Salisbury thought he saw a way out: Ibid.
281 "No word was . . .": Ibid.
281 "If it is . . .": Macmillan, *Diaries*, July 23, 1954.
282 "stag-party": Colville, *The Fringes of Power*, August 1954.
282 "air of crisis": Ibid.
282 "new event": Cabinet Secretary's Notebooks, July 26, 1954, CAB 195/12.
282 said he was satisfied: Ibid.
283 "I don't regret . . .": Ibid.
283 happy to assent: Ibid.
283 Molotov expressed astonishment: The Prime Minister, Mr. W. Churchill, From V. M. Molotov, July 31, 1954, PREM 11/669.
284 fresh set of arguments for bilateral talks: Two Power Meeting with Soviet Government, Note by the Prime Minister, August 3, 1954, PREM 11/669.
284 Salisbury's response: Lord Salisbury to Prime Minister Churchill, August 20, 1954, PREM 11/669.
285 "In peace or war . . .": Prime Minister Churchill to Lord Salisbury, August 21, 1954, PREM 11/669.
285 French would not be ratifying: WSC to Dulles, August 24, 1954, *FRUS: 1952–1954*, vol. 5, part 1.
285 wrote to Eden: WSC to Anthony Eden, August 24, 1954, quoted in Gilbert, *Churchill*, vol. 8, 1050.
286 "It has gone . . .": WSC to Clementine Churchill, August 25, 1954, in Soames, ed., *Winston and Clementine*.

XIX. The "R" Word

287 "This is to . . .": WSC, November 30, 1954, quoted in "On the Occasion of Churchill's 80th Birthday," The Churchill Society, London.

288 "just the moment . . .": Lord Salisbury to Paul Emrys Evans, November 28, 1954, PEE Papers.

288 "That is something . . .": *The Times*, December 1, 1954.

289 "It is in . . .": Prime Minister Churchill to President Eisenhower, December 7, 1954, in Boyle, ed., *The Churchill-Eisenhower Correspondence*, 180.

289 "play tough": President Eisenhower to Prime Minister Churchill, December 14, 1954, in *The Papers of Dwight David Eisenhower*, vol. 15, 1444.

290 general election in the spring: Anthony Eden, diary entry, December 21, 1954, quoted in Rhodes James, *Anthony Eden*, 392.

290 told Eden on the phone: Eden, *Clarissa Eden*, 181.

290 Clementine Churchill expressed alarm: Anthony Eden, diary entry, December 21, 1954, quoted in Rhodes James, *Anthony Eden*, 392.

290 "What do you . . .": Ibid.

291 "one against the . . .": Macmillan, *Diaries*, October 10, 1954.

291 "I know you . . .": Quoted in Anthony Seldon, *Churchill's Indian Summer* (London: Hodder, 1981), 51.

291 no one contradicted: Anthony Eden, diary entry, December 22, 1954, quoted in Rhodes James, *Anthony Eden*, 393.

291 "But if this . . .": Quoted in Seldon, *Churchill's Indian Summer*, 51.

291 "in despair": Macmillan, *Diaries*, December 22, 1954.

291 Stuart opined: Anthony Eden, diary entry, December 22, 1954, quoted in Rhodes James, *Anthony Eden*, 393.

291 until he died: Macmillan, *Diaries*, December 22, 1954.

292 reconvened at the home: Seldon, *Churchill's Indian Summer*, 51.

292 rise to the great occasion: Norman Brook quoted in Gilbert, *Churchill*, vol. 8, 1084.

292 Colville admitted: Colville, *The Fringes of Power*, March 29, 1955.

292 "tired of it all": Ibid.

293 sending for a calendar: Anthony Eden, diary entry, February 1, 1955, quoted in Rhodes James, *Anthony Eden*, 396.

293 tell the Queen quietly: Clarissa Eden, diary entry, February 2, 1955, quoted in Eden, *Clarissa Eden*, 182.

293 convinced that he finally: Clarissa Eden, diary entry, February 14, 1955, ibid., 184.

293 he told Moran: Moran, *Diaries*, February 21, 1955.

294 "What ought we . . .": WSC, March 1, 1955, quoted in *New York Times*, March 2, 1955.

294 "a very fine . . .": Moran, *Diaries*, March 1, 1955.

294 "It is absolutely . . .": *Hansard*, March 2, 1955.

294 trembled for fear: Macmillan, *Diaries*, March 2, 1955.

294 Soames chewed his thumb: Ibid.

295 "slid successfully past . . .": Ibid.

295 "At least not yet": Moran, *Diaries*, March 3, 1955.

296 go to Sicily: Clarissa Eden, diary entry, March 8, 1955, quoted in Eden, *Clarissa Eden*, 194.

296 Eden expressed confidence: Macmillan, *Diaries*, March 9, 1955.

296 Clementine gave Clarissa a tour: Clarissa Eden, diary entry, March 10, 1955, quoted in Eden, *Clarissa Eden*, 194.

296 tested the new Rolls-Royce: Colville, *The Fringes of Power*, March 29, 1955.

296 changed his mind: Ibid.

297 "lay plans for . . .": WSC to Anthony Eden, March 12, 1955, quoted in Rhodes James, *Anthony Eden*, 401.

297 "the first time": Ibid.

297 "The magnitude of . . .": Ibid.

297 "full crisis": Clarissa Eden, diary entry, March 11, 1955, quoted in Eden, *Clarissa Eden*, 194.

298 "would justify the . . .": Anthony Eden to WSC, March 12, 1955, quoted in Rhodes James, *Anthony Eden*, 402.

298 might not be able to stand the strain: Macmillan, *Diaries*, March 13, 1955.

299 "I don't think . . .": Cabinet Secretary's Notebooks, March 14, 1955, CAB 195/13.

299 "Does that mean . . .": Macmillan, *Diaries*, March 14, 1955.

299 "This is a . . .": Cabinet Secretary's Notebooks, March 14, 1955, CAB 195/13.

299 "All this is very . . .": Macmillan, *Diaries*, March 14, 1955.

300 "new offer": Cabinet Secretary's Notebooks, March 14, 1955, CAB 195/13.

300 "It isn't against . . .": Clarissa Eden, diary entry, March 14, 1955, quoted in Eden, *Clarissa Eden*, 197.

300 Aldrich delivered: WSC to Clementine Churchill, March 15, 1955, in Soames, ed., *Winston and Clementine.*

300 insisted on going back: Ibid.

301 "a gesture of . . .": Macmillan, *Diaries*, March 16, 1955.

301 "It's the first . . .": Quoted in Soames, *Clementine Churchill*, 493.

301 no "next time": Ibid., 492.

301 "But Ike won't . . .": Quoted in Bonham Carter, *Daring to Hope*, April 1, 1955, 147.

302 gaffed: Quoted in Seldon, *Churchill's Indian Summer*, 53.

302 landed: Colville, *The Fringes of Power*, March 29, 1955.

302 just to gratify: Ibid.

302 the young sovereign said no: Ibid., 661.

303 Colville sent word: Ibid., March 29, 1955.

303 inquired how his niece: Eden, *Clarissa Eden*, 198.

303 scene in Cabinet Room: Butler, *The Art of the Possible*, 176.

304 "to die in . . .": Winston S. Churchill, *Marlborough: His Life and Times*, vol. 2 (London: Harrap, 1936), 1036.

305 "I don't believe . . .": Colville, *The Fringes of Power*, 662.

305 "I have decided . . .": Cabinet Secretary's Notebooks, April 5, 1955, CAB 195/13.

INDEX